Ajay Kumar Goswami

Spectrophotometric Determination of Rare Earth Elements

Also of Interest

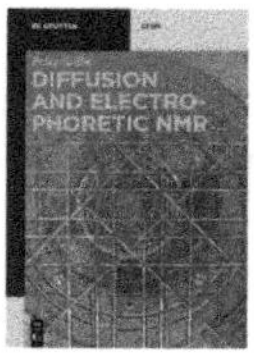

Diffusion and Electrophoretic NMR
Peter Stilbs, 2019
ISBN 978-3-11-055152-5, e-ISBN 978-3-11-055153-2

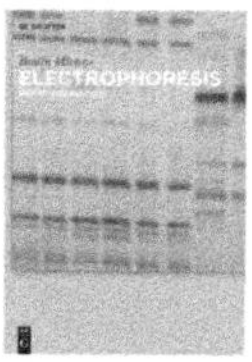

Electrophoresis.
Theory and Practice
Budin Michov, 2020
ISBN 978-3-11-033071-7, e-ISBN 978-3-11-033075-5

Atomic Emission Spectrometry.
AES - Spark, Arc, Laser Excitation
Heinz-Gerd Joosten, Alfred Golloch, Jörg Flock, Susan Killewald, 2020
ISBN 978-3-11-052768-1, e-ISBN 978-3-11-052969-2

Chemical Analysis in Cultural Heritage
Volume 1
Luigia Sabbatini, Inez Dorothé van der Werf (Eds.), 2020
ISBN 978-3-11-045641-7, e-ISBN 978-3-11-045753-7

Ajay Kumar Goswami

Spectrophotometric Determination of Rare Earth Elements

Reagents and Methods

DE GRUYTER

Author
Prof. Ajay Kumar Goswami
Subhash Nagar
132 Road No. 5
Udaipur-313001
India
akumargoswami56@gmail.com

ISBN 978-3-11-161483-0
e-ISBN (PDF) 978-3-11-161500-4
e-ISBN (EPUB) 978-3-11-161539-4

Library of Congress Cataloging-in-Publication Data
A CIP catalog record for this book has been applied for at the Library of Congress.

Bibliographic information published by the Deutsche Nationalbibliothek
The Deutsche Nationalbibliothek lists this publication in the Deutsche Nationalbibliografie; detailed bibliographic data are available on the Internet at http://dnb.dnb.de.

Cover image: Ivan-balvan / iStock / Getty Images Plus
Typesetting: VTeX UAB, Lithuania

www.degruyterbrill.com
Questions about General Product Safety Regulation:
productsafety@degruyterbrill.com

Contents

1 Rare earth elements – an introduction

The rare earth elements we see on Earth today originated in a supernova explosion about ten billion years ago. Almost one hundredth of one percent of the mass of all existing elements in Earth's crust is accounted for by rare earth elements. While this may sound small, the quantity of all precious metals is yet smaller. The greatest difficulty extracting them is that they have not separated into elemental minerals despite having been subjected to rigorous geological processes. Moreover, it is interesting to know that the relative abundance of the rare earth elements in chondritic meteorites show their overall abundance in the universe, and consequently this make them important to study for an explicit understanding of the genesis of chemical elements.

The history of discovery of these elements is the origin of the name rare earth They are never found as free metals in the Earth's crust, even though pure minerals of free metals do exist. Since they are found only as oxides, chemists of 18th and 19th centuries had difficulty separating them from each other. The ancient Greeks defined 'earths' as those materials that could not be decomposed further by being subjected to heat, and these oxides fit that definition. The word 'rare' with implied the difficulty in obtaining the pure elements and not to their relative abundance in the Earth's crust. In fact, some rare earths exist in greater abundance than lead.

Fourteen elements, from cerium to lutetium, alternatively named lanthanons, lanthanides and rare earths are characterized by completely filled antepenultimate 4f energy levels. They are also called f-block elements. Since the present book is primarily devoted to spectrophotometric methods for the determination of rare earths, lanthanides, or lanthanons, it is pertinent to consider this aspect in detail. Moreover, the preceding section offers introductory discussion on rare earth elements that apprises readers of the basic properties and application of some rare earth elements. The global trends that will shape the economies of the next decade necessarily include rare earth metals. They are used in many high-tech products from smart phones to cars, electrical appliances and green technologies like hybrid and electric vehicles to wind turbines, to name a few.

1.1 Rare earths and their key attributes

An interesting feature of rare earths or lanthanide is the occurrence of lanthanide contraction. This phenomenon is a steady decrease in atomic size as well as ionic size with increasing atomic number. A major cause of this contraction is the electrostatic effect of increasing nuclear change that is perfectly screened by the 4f electrons. This is comparable to the smaller contraction observed through each of the d-block elements of the transition series. Furthermore, the partial impact is due to the relativistic effect that influences the shielding characteristics of inner shell electrons [1].

https://doi.org/10.1515/9783111615004-001

1.2 Color and spectra

There is striking difference in spectroscopic properties of lanthanides compared to d-block elements. This is due to 4f electrons that are shielded from external fields by the overlying $5s^2$ and $5p^6$ shells. Thus, the states arising from $4f^n$ configurations tend to remain almost invariant for a given ion. The origin of color is due to the fact that light of a particular wavelength is absorbed corresponding to particular electronic transition. In general, lanthanides show absorption in the visible or near UV regions of the spectrum except Lu^{3+} which has full f shells. The colors in lanthanide arise due to f–f transition. As the f orbitals are deeply located inside the atom, they are in fact largely shielded and, as such, due to complex formation the absorption spectra are sharp. This is different from d–d spectra which are broad. This implies that the nature of ligands little effects the spectra compared to d-elements. This adds to the fact that these spectra are useful both for qualitative detection and quantitative determination. Furthermore, ions like Ce^{3+} and Yb^{3+} show strong absorption in UV region and thus are colorless. It is possibility that CT spectra are due to ligand–metal transfer, while metal in high-oxidation state, e. g., strong color in Ce^{4+} solution, is due more to CT than f–f transitions. Table 1.1 formulates colors in both the solid state and in aqueous solutions for lanthanides (M^{3+}) ions.

Table 1.1: Colors vs f-electrons.

(Element) M^{3+} ion	Number of 4f electrons	Color
La^{3+}	0	Colorless
Ce^{3+}	1	Colorless
Pr^{3+}	2	Green
Nd^{3+}	3	Lilac
Pm^{3+}	4	Pink
Sm^{3+}	5	Yellow
$E4^{3+}$	6	Pale pink
Gd^{3+}	7	Colorless
Lu^{3+}	14	Colorless
Yb^{3+}	13	Colorless
Tm^{3+}	12	Very pale green
Ey^{3+}	11	Pink
Ho^{3+}	10	Pale yellow
Dy^{3+}	9	Yellow
Tb^{3+}	8	Pale pink

1.3 Application of rare earth elements (REEs)

Rare earths comprise an extremely important material in the economy, with rapidly developing technologies worldwide. Industrial sectors such as automobiles, renewable energy, nuclear power, and catalysis are substantially impacted by their application. To keep it as brief as possible, the following description is focused on various applications of rare earths in industry or elsewhere.

(a) **Permanent magnets**

The name denotes that they exhibit permanent magnetic forces without external influence like electrical voltage. In this regard, the strongest magnets consist of a high ratio of rare earths. These are especially used where high magnetic power is required in confined volumes. Wind power, electro mobility, and other ecofriendly technologies warrant the use of permanent magnets. Neodymium, praseodymium, dysprosium, terbium and samarium are applied for such applications.

(b) **Catalysts**

Application of catalysts in pharmaceutical, chemical, or other similar industries is not new. The best-known application in automobiles is as the catalytic converter that reduces pollutants to nonpolluting residues. Neodymium, lanthanum and praseodymium are rare earths used as catalysts, and refining crude oil also uses catalysts. In particular, cerium, samarium and ytterbium are used as catalysts.

(c) **Polishes**

Polishing is used to make mirrors, automobile wind screens, and similar components meet the requirements, and it has been shown that polishes containing rare earth elements are effective for removing scratches from such parts. What's more, cerium and neodymium are often the best choice in this regard.

(d) **Batteries**

Since batteries are fundamental devices for energy storage, their application is necessary both in hybrid and fully electrical vehicles. Alloys containing rare earths are used in electrodes of these batteries. Lanthanum and neodymium are used in batteries.

(e) **Glass**

Rare earths are essential material for special glasses used in space travel as well as in touch displays in cell phones. They are also used to color in decolor glasses as well as in mirrors. In ceramic materials, various rare earths are used as colorants and polishing agents, while they can even increase the data-transmission performance of fiber-optic materials.. Cerium, neodymium, praseodymium and holmium are useful for this application.

(f) **Alloys**

Since centuries, alloys have been part of human civilization. Today, rare earths play an important role in alloy industry. Moreover, from aeroplanes to aircraft engines, rare earths play pivotal roles in aerospace. Ytterbium, dysprosium, terbium, praseodymium, and neodymium are some rare earths used for alloying purposes.

(g) **Ceramics**

Scandium, yttrium and cerium are used in ceramic products. The important metallic combinations in ceramics for improving resistance to heat and physical stress include rare earths. Manufacturing of medical implants and fuel cells incorporate rare earths.

(h) **Lighting**
Some of the applications of rare earths include lighting, fluorescent tubes, and modern energy saving lamps and LED lights. Using rare earth metals ensures, for example, that blue light emitted by light emitting diode (LED) is also converted to white, green, and red light. They are also used in devices with picture tubes and tablets and also in smart phones to produce red pixels. A number of rare earth elements, e. g., dysprosium, erbium, europium, lanthanum, luteum, gadolinium, among others, are applied for such use.

(i) **Pigments**
Coloring agents unlike dyes remain solids when they are used to color objects such as ceramics, glass, enamels, etc. In fact, this area is one where rare earths play an important role. Glass coloring was the first application of rare earths, and praseodymium is used to causer yellow, blue and green coloration. Neodymium imparts shades of pink to violet, and cerium produces an orange color when used in enamels. A number of rare earths are used for such applications, including erbium, europium, gadolinium, holmium, neodymium, samarium and praseodymium.

(j) **Wind power**
Permanent magnets made out of rare earths are very important components of wind turbines which convert rotational energy to electrical energy. The increasing demand for wind energy has opened up major application of rare earths, especially useful in this regards are dysprosium, neodymium, praseodymium and terbium.

(k) **Nuclear power**
Demand for rare earths has significantly increased due to their application in the production of nuclear power. Some of them are used in the controlling rods of reactors, which regulates the nuclear reactor. Examples of rare earths used in reactors are dysprosium, europium, samarium, holmium and gadolinium.

(e) **Water electrolysis**
The green energy race and sustainable exploration demands green hydrogen production, which is produced by electrolysis of water. Various rate earth elements are used as electrolytes, and this increases the demand for rare earth elements. Scandium, yttrium, cerium, gadolinium and lanthanum are some rare earths useful in this application.

(m) **Electromobility**
In recent years, climate change and the fight against it have urgently warranted alternative sources and technologies for transportation not based on petroleum. The electromobility in view of this has given rise to EV models, and consequently rare earths are used in magnets found in drive motors that convert electricity to kinetic energy. Rare earths further help to provide these magnets with the required weight-to-power ratio, thus becoming more popular. Terbium, dysprosium, neodymium, praseodymium and terbium are a few of the rare earths useful for this purpose.

(n) **Aviation**

This is another area where rare earth elements find application. A US Congress report mentions that a single F-35 fighter jet plane contains about 417 kg of rare earths. Apart from their use in electronics and landing-gear motors, their alloys are also used to make wings. Furthermore, tire production using them as catalysts is the rule. Samarium, scandium, europium, etc. are some rare earths useful in the aviation industry.

(o) **Lasers**

It is known well that industry uses lasers to cut, shape, weld or coat materials. More broadly, even diagnostic tools, dermatological procedures, dentistry and surgery make extensive use of lasers. Rare earths used in lasers include terbium, neodymium, holmium, thulium and erbium.

(p) **Medicine**

Rare earths, due to their unique properties such as radiation emission and magnetism, are very useful for a number of therapeutic and diagnostic applications in modern medicine. Zhang et al. [2] confirmed the antifungal and antibacterial activities of some rare earth elements. Gadolinium has been used in chelated form as a contrast in magnetic resonance imaging (MRI) as reported by Raju et al. [3]. In sum, these elements, since they improve image quality, are useful in X-ray intensifier foils and contrast agents. Gadolinium, luteum, terbium, yttrium, neodymium and erbium are some elements employed of this application.

(q) **Semiconductors**

The doping of rare earths numerous applications in semiconductors and electronic components like LEDs, microprocessors, solar cells or lasers. A host of rare earths are used for such applications, including cerium, erbium, lanthanum, neodymium, praseodymium, samarium, scandium, yttrium and ytterbium.

Bibliography

[1] Pitzer, K. S., Accounts of Chemical Research 12 (1979), 271.

[2] Zhang, H., Feng, J., Zhu, W. F., Liu, C. Q., Gu, J. H., Bactriostatic effect of cerium-humic acid complex an experimental study, Biological Trace Element Research 73(1) (2000), 29–36.

[3] Raju, C. S. K., Cossmer, A., Scharf, H., Panne, U., Luck, D., Speciation of gadolinium based MRI contrast agents in environmental water samples using hydrophilic interaction chromatography hyphenated with inductively coupled plasma mass spectrometry, Journal of Analytical Atomic Spectrometry 25 (2010), 55–61.

2 Spectrophotometric methods principles and practice

We know that every chemical compound absorbs, transmits or reflects electromagnetic radiation (light) over a certain wavelength range. Spectrophotometry is one method that measures how much a chemical substance absorbs light by measuring the intensity of beam of light that passes through a sample solution of the substance in question. This implies that spectrophotometry measures the amount of a known chemical substance and is one of the important tools for quantitative analysis of transition metal complexes and rare earth metals. The technique is one of the most versatile methods for quantitative analysis of not only metal complexes but also for clinical, biochemical, pharmaceutical or even metallurgical samples. These methods are characterized by versatility, sensitivity and precision, and thus are the most commonly used reliable methods for the determination of majority of elements. Over a wide range of concentrations, the methods are applicable varying from macro quantities to trace determination in the range of 10^{-6}–10^{-8} M, making them the most precise methods for chemical estimation. The other advantage of spectrophotometry is its easy and economic instrumentation as compared to other methods for chemical analysis. Thus, there is no doubt that spectrophotometric methods play a pivotal role in the area of chemical analysis.

2.1 Basics of spectrophotometry

The spectrophotometer is an instrument that measures the amount of photons or the intensity of light absorbed after it has travelled through a sample solution. With such an instrument, the quantity of a known chemical substance or the concentration can be determined by measurement of the intensity of light. There are two basic modes of spectrophotometry based on the region or wave length of light source, namely,

(i) UV-visible spectrophotometer: This uses light over the UV range (700–1,500 nm) and visible region (400–700 nm).
(ii) IR spectrophotometer: This uses light in the infrared region (i. e., 700–15,000 nm)

2.2 Fundamentals of spectrophotometry

Spectrophotometric measurements are based on a simple principle, namely, Lambert Beer Law or Beer's Law. The law states that there is a linear relationship between the absorbance and the concentration of any sample. Thus, this law can be applied only when the relationship between the concentration and absorbance is linear. It is expressed as

$$A = \varepsilon lc$$

https://doi.org/10.1515/9783111615004-002

where
- A is the absorbance
- ε is the extinction coefficient or molar absorptivity
- l is the path length
- c is the concentration.

A standard spectrophotometer uses a cuvette of 1 cm in width, while l is always assumed to equal 1 cm. As we know the absorption, ε and the path length of cuvette (1 cm), we can calculate the concentration $[c]$ of the sample.

2.3 Terms used throughout this book

Spectrophotometry is one of the most useful technique for determining trace amounts of metals in micro or macro quantitative measurements. It was developed by T. S. West using the UVVis region of the electromagnetic spectrum. This section deals with the method and brief defines the terms used in the book. This can help any reader to understand the book more clearly.

Choice of wavelength
Selecting a suitable wavelength is very important in spectrophotometry. This is done by selecting the wavelength of absorbance in a region where the molar absorptivity (ϵ) changes quickly with the wavelength. So to say a wavelength is selected for a colored solution, where absorbance is maximum and constant with a small change in wavelength.

Color development
The color of any solution is explained by the fact that a solution appears colored when it emits or absorbs a particular range of the radiation in the visible region. Molecular structure of a constituent and its color are closely related to each other. Thus, this is the concept behind the measurement of the concentration by calculating the extent of absorbance. In metal complexes, the instant of color development is not always conant, and it may take some time before the solution attains maximum color intensity.

Deviation from Beer's Law
This is basic law governing spectrophotometric determination. This is expressed as a plot of absorbance versus concentration, and a straight line passing through origin is obtained. In a dissolved state, it covers a wide range of concentrations. The presence of some small number of electrolytes normally does not affect the absorbance. Yet, in the presence of a large amount of such electrolytes, a shift in maximum absorption may be observed that changes the molar extinction coefficient. Beer's law is the basis of quantitative application of UV-Vis spectrophotometry.

Sensitivity

The minimum amount of concentration of an element that can be determined by an analytical method is called the sensitivity. The numerical value of this is determined by the molar absorption coefficient and is expressed as

$$\varepsilon = \frac{A}{Cl}$$

where symbols represent the same terms as earlier stated.

The general criterion for describing sensitivity are as follows:

- Low sensitivity $\varepsilon < 2 \times 10^4$ L/mol/cm
- Moderate sensitivity $\varepsilon = 2–6 \times 10^4$ L/mol/cm
- High sensitivity $\varepsilon > 6 \times 10^4$ L/mol/cm.

Precision and accuracy

The term precision denotes the reproducibility of the analytical results. In fact, the degree of proximity of the obtained results in the true value. It ranges from 0.2 to 0.5 % in any differential technique and from 0.5 to 2 % in general.

Selectivity

This is based on the characteristic of an analytical reagent used for complexation. A selective reagent is one that reacts only with a small number of elements, whereas a specific reagent is one that reacts only with one element under definite conditions. Thus, this depends on the color reaction of the metal-ligand (reagent), the pH and interfering ions, in addition to a few other conditions.

Interfering species and diverse

Application of any method is more successful if there is no interference of diverse ions and cations, similar to the ion in question. The tolerance limits of such ions are established at these particular concentrations, where there is not more than ±2.0 % of change in absorbance of the analyte in question.

Stoichiometry

The technique of spectrophotometry is best applied to determine the stoichiometry of any complex. There are three commonly used methods based on mole ration viz.,

i) Job's method of continuous variation
ii) the Slope ratio method
iii) Turner Anderson's method

Readers may consult any standard book for details.

Stability constant of complex

The stability of a metal complex formed determines the selectivity of that method. Thus, stability constants play an important role in understanding the behavior of a chelate; they are dependent on many factors like the size of the ring, the basic strength of chelating molecule and the atomic nature of the ion.

The standard addition method

An ideal calibration standard usually approximates the composition of the analyte in question with respect to both the concentration of analyte and other species in the sample. This implies minimum effects of the various components in the absorbance measurement.

2.4 Spectrophotometric reagents: a brief introduction

In view of the fact that most spectrophotometric methods are based on the use of organic reagents [1–10], it is warranted to present introductory details of the various reagents used. The literature contains many reviews of organic reagents, including those by this author and his co-researchers for various metals [11–13]. Moreover, a series of spectrophotometric methods for metals like Cu, Fe, Ni, Co, V, Cr, Mn, Pd and Pt has been published by the author [14–17]. Thus, it becomes complicated to review the scenario of reagents and methods, and so only a brief introduction of some common reagents is presented here.

2.4.1 Azo reagents

Azo dyes offer sensitive color reactions with metals, and this group includes reagents like 1-(2-pyridylazo)-2-naphthol (PAN), 4-(z0pyridylazo) resorcinol (PAR), 5-Br-PADAP, chlorophosphanazo III and arsenazo III, among others. For the determination of metals [18, 19], pyridylazo and thiozolylazo reagents constitute an important class of reagents.

1-(2-pyridylazo)-2-naphthol (PAN)

PAN can exist in three forms, depending on pH of the solution. At pH ≤ 2, it shows a water soluble yellow–green protonated H_2R^+ form. In the pH range of 3–11, there exists as a neutral HR molecule, which is soluble in organic solvents to give yellow color. Above pH ≥ 11, there exists a water-soluble red R^- ion. The structure of the reagents can be shown in figure 2.1.

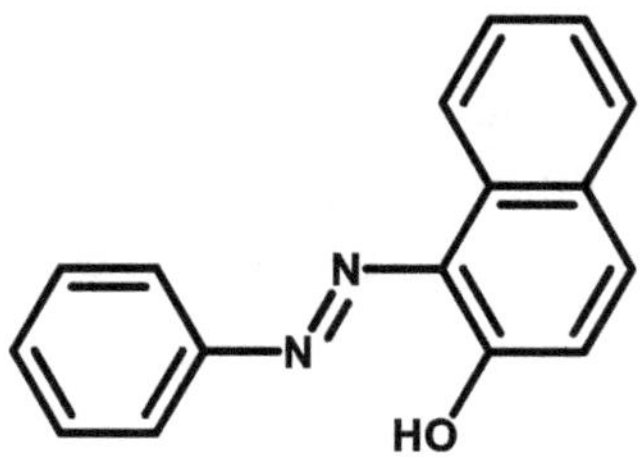

Figure 2.1: 1-(2-pyridylazo)-2-naphthol.

PAN acts as a tertradentate ligand that has donor sites with metal through hydroxyl oxygen, pyridine nitrogen atom and one of the azo nitrogen. Normally, the molar absorptivities of PAN complexes range 2–6×10^4. The advantages with PAN is that the absorption maxima λ_{max} of the metal complexes are much different than the reagent, rendering the determination easier to make.

4-(2-pyridylazo) resorcinol (PAR)

The structure of PAR is shown in Figure 2.2.

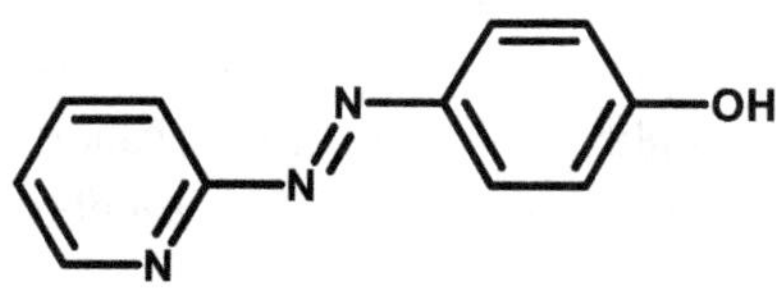

Figure 2.2: 4-(2-pyridylazo) resorcinol.

The reagent is water soluble and forms water-soluble metal complexes so that spectrophotometry can be performed in an aqueous medium. In 0.5–0.05 m, H_2SO_4 PAR forms a complex with Zr, Pd and Ti (III) ions, while, on the other hand, in pH 3–6 it produces a color reaction with a variety of ions like Zn, Co, Na, Hg, V, among others. The color of the complexes with PAR is usually read or violet.

It is of interest to note that metal complexes of PAR are extractable into chloroform in the presence of diphenylguanidine [20]. Highly sensitive methods using bromo and chloro derivatives of PAR are also significantly important. The one that is most widely used is 2-(5-bromo-2-pyridylazo)-5-dimethylaminophenol (5-Br-PADAP), which has become a useful reagent for Cd, Co, Ni, etc. The structures such reagents has been expressed in figure 2.3.

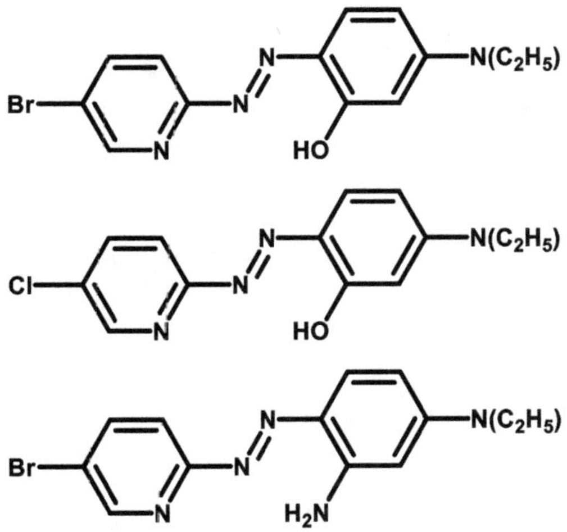

Figure 2.3: Derivatives of 4-(2-pyridylazo) resorcinol.

Thiazoleazo compounds

These compounds have similar properties as PAN and PAR for application as spectrophotometric reagents. Since their stability is lower, their selectivity as reagents offer big advantages as some of the following examples shown in figure 2.4.

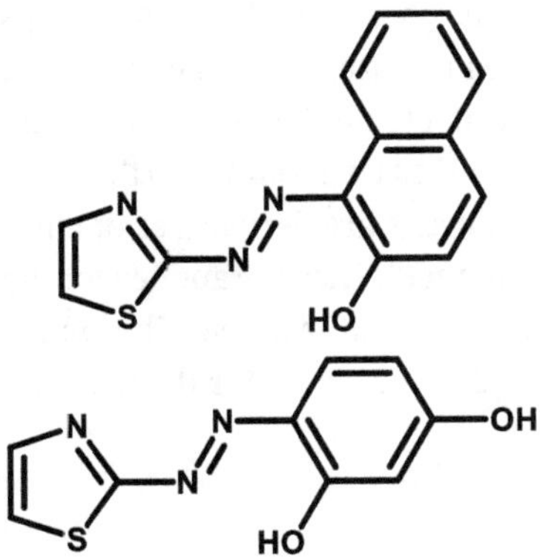

Figure 2.4: Thiazoleazo compounds.

A review on thiazylazo reagents and their application as spectrophotometric reagent has been published [21].

Arsenazo compounds

Arsenazo I (Fig. 2.5) and Thoron I are the first reagents of this group and are characterized by the presence of an arsenic acid group at ortho position to azo group. The –OH group also is ortho to the azo group. The structures can be shown to be as follows:

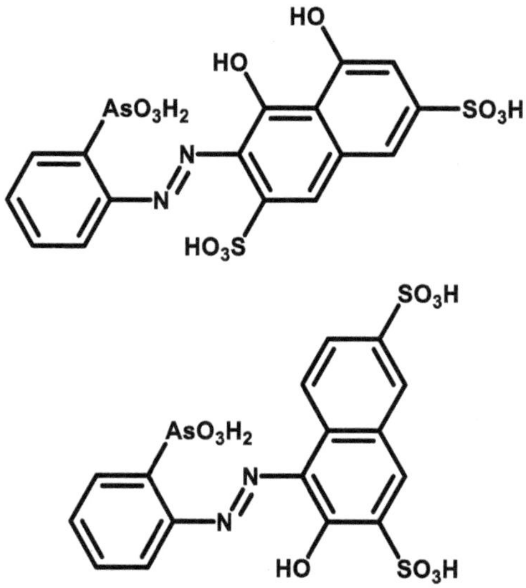

Figure 2.5: Arsenazo compounds.

Stable complexes are formed due to the presence of ASO_3H_2 even in fairly acidic medium. Azo group being a good donor forms complex easily and OH group further forms a second ring with the metal. Not only are the complexes formed stabilized, but it also improves the colors obtained. However, the presence of a sulphonic group helps in making the complex water soluble, making the entire measurement easier. It should be mentioned that hardly any sensitive reagent was available until arsenazo I was introduced, particularly for determination of Th, Zr, Hf and U as well as rare earths.

It will be helpful to note that Arxenzo I and its analogues are derivatives of chromotropic acid or 1,8-dihydroxy hapthalene-3,6-disulphonic acid. They are applied for the determination of the rare earths Th, U, Zr, Ti Nb etc. as a reagent, whereas Thoron I is comparatively less sensitive and is most commonly used as a reagent for the determination of Li, U, Zr, Be and Th.

Arsenazo III

This is a very good bisazo dye which is based on chromotropic acid and O-amino phenylarsonic acid. It is readily soluble in alkaline solutions as well as moderately soluble in neutral and acidic solutions. The reagent decomposes with strong oxidizing agent like H_2O_2, Cl_2 and Br_2 and reducing agents like $TiCl_3$ also cause decomposition. The Arsenazo III has a purplish-red color from 10 m HCl to pH 4 and at higher pH value attains a blue violet color. The structure is represented in Fig. 2.6.

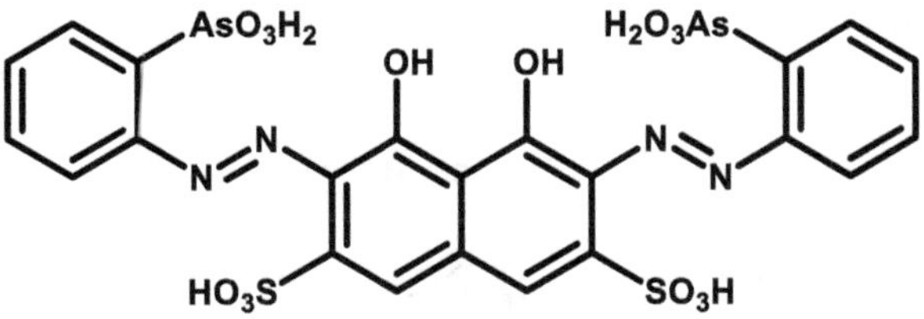

Figure 2.6: Arsenazo III.

The reagent reacts in a strong acidic medium 1–10 m HCl, for example, with Th, Zr, Hf and U (iv), and with a good molar absorptivity value of 10^5. On the other hand, the reagent reacts with U (vi), Sc, Fe (iii) and rare earth in pH range 1–4. In such conditions, the sensitivity of this color reaction reduces to $\varepsilon = 5\times10^4$. Thus, to avoid difficulties with hydrolysis of some ions like Zr, Th and U, a strongly acid medium can be recommended, which also enhances selectivity with the reagent.

The reagent absorbs at λ_{max} 520–530 nm, but the metal complexes have a slightly higher λ_{max} of 655–665 nm. If we look into the mechanistic side of Arsenazo III, only one side of the symmetrical molecule is associated with the formation of complexes with the metal: the metal coordinate through the nitrogen atom of the azo group, the oxygen atom of arsonic group and the oxygen atom of the hydroxyl group. This causes distortion of the symmetry of the molecule producing two neighboring absorption maxima in the visible region for the Arsenazo III–metal complexes. In concentrated H_2SO_4, it absorbs at 675 nm, and it can be determined that other similar reagents are azo-dyes containing phosphonic acid groups. An example is chorophosphanazo III (Fig. 2.7), which is useful for determination of Zr, Ti, U, Sc, Ca and Sr [22].

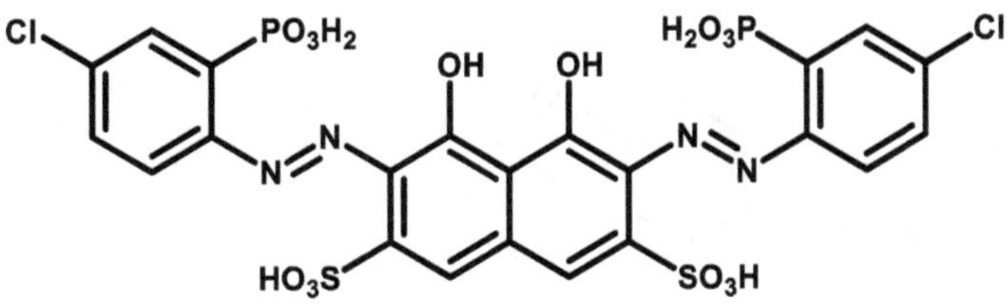

Figure 2.7: Chlorophosphanazo III.

Miscellaneous Azo Compounds

There are numerous similar azo compounds applied to spectrophotometric determination of metals like Nb, Zr, Mo, Cu, V, Al in an acid medium: a list of such reagents is provided in figures 2.8–2.11.

(i) Sulphochlorophenol S

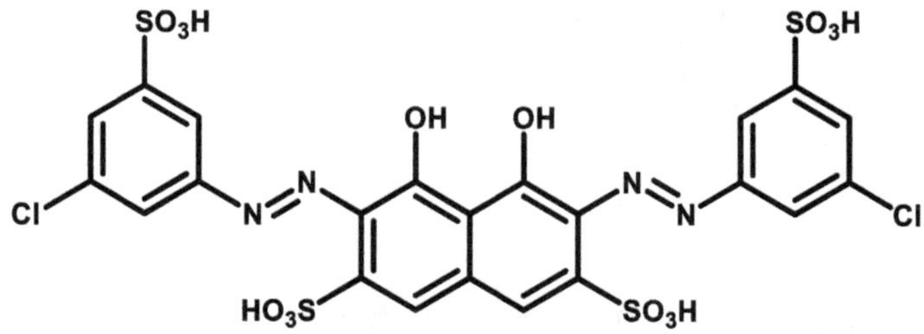

Figure 2.8: Sulphochlorophenol S.

(ii) Sulphonitro phenol M

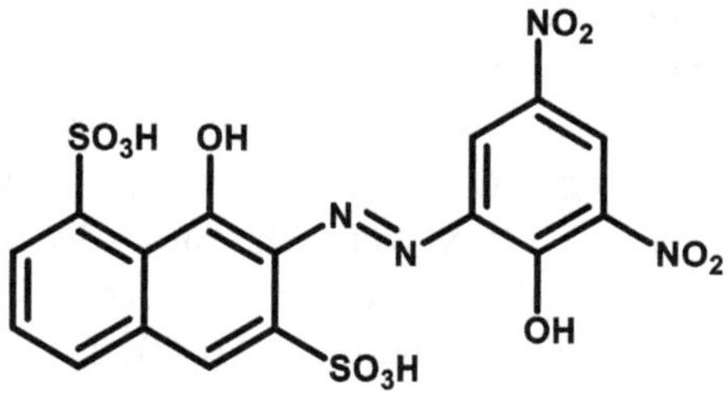

Figure 2.9: Sulphonitro phenol M.

(iii) Chromotrope 2B

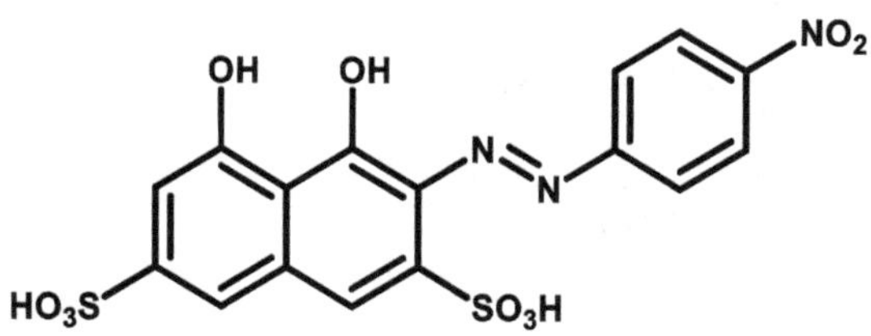

Figure 2.10: Chromotrope 2B.

(iv) 2-(4, sulphophenylazo) chromotropic acid (SPADNS)

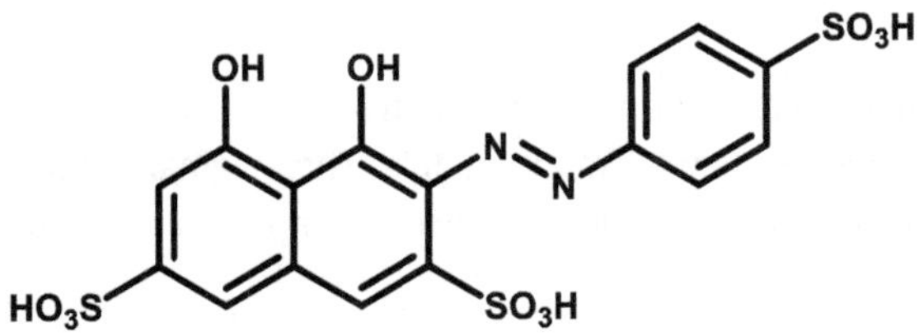

Figure 2.11: 2-(4, sulphophenylazo) chromotropic acid.

Excellent reviews of research involving 8-hydroxyquinoline azo derivatives, pyrocatecholazo derivatives and o-thiazo derivatives of p-cresol and 2-naphthol [23–25] are available for the interested reader.

2.4.2 Triphenylmethane reagents

The presence of p-quinonoid rings in triphenylmethane class of reagents makes them intensely colored and thus also extremely good reagents for spectrophotometry of number of metals. This is a vast area with voluminous details so few examples are given to make the reader aware of this class.

Pyrocatechol violet

Water and aqueous-ethanol soluble pyrocatechol is a commonly used chelating reagent for metal determination. In the pH 1–8 range, the aqueous solution of this reagent is yellow in color, but it changes to violet with measuring pH. Almost all chelates formed with metals with the reagent are blue, under weakly acidic to weakly basic conditions. Figure 2.12 shows the structure of reagent.

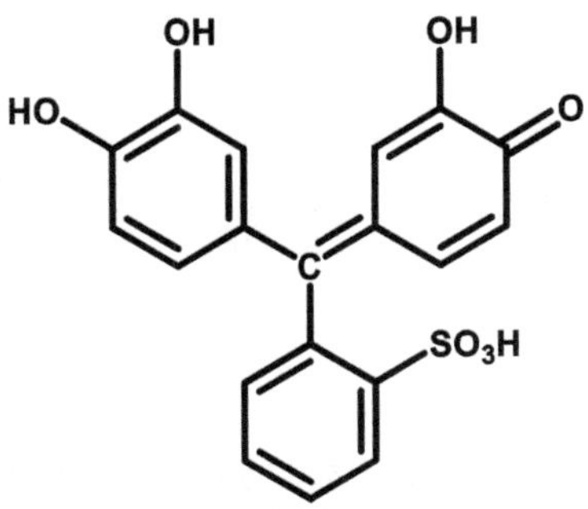

Figure 2.12: Pyrocatechol violet.

Eriochrome Cyanin R (ECR)

This reagent has additionally a carboxylate group besides hydroxyl and sulphonate groups. In weakly acidic solutions, ECR reacts with many metals. The original yellow-orange color of the reagent turns to violet or blue as this forms complexes with any metal. (Fig. 2.13)

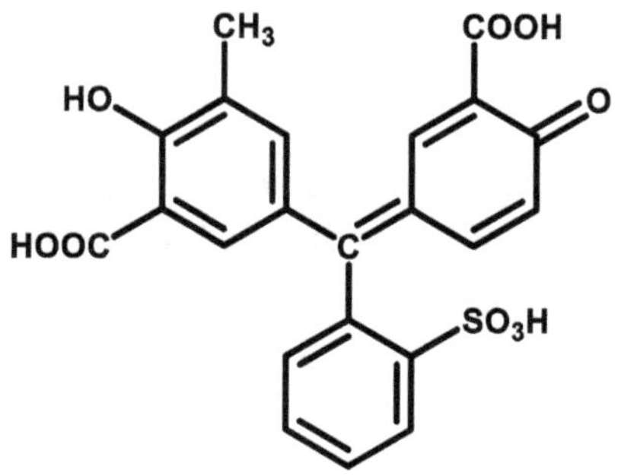

Figure 2.13: Eriochrome Cyanin R.

Chrome Azurol S (CAS)

The reagent CAS reacts with many metal in a weakly acidic or neutral medium. It also changes to blue or violet when it forms complexes with metal. Figure 2.14 shows the structure of CAS.

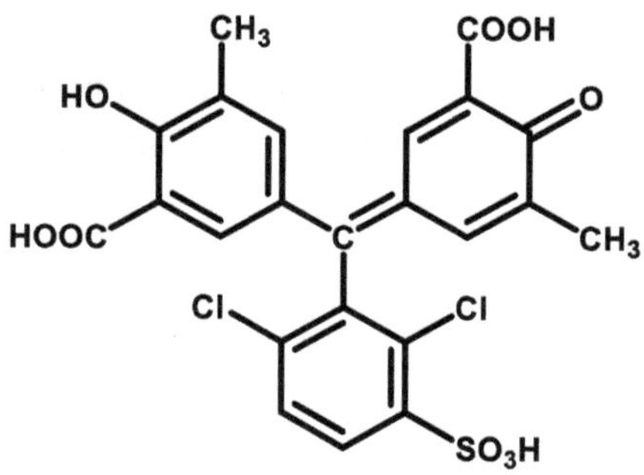

Figure 2.14: Chrome Azurol S.

Xylenol orange

Xylenol orange and its analogue, methylthymol blue, are characterized by the presence of an iminodiacetic acid group, a structural feature they share with EDTA. Both xylenol orange and methylthymol blue are good chromogenic reagents for Sc, Ti and Zrsince these are easily hydrolysable. The values of ε exceeds 10^5. Structures and represented in Figs. 2.14 and 2.15.

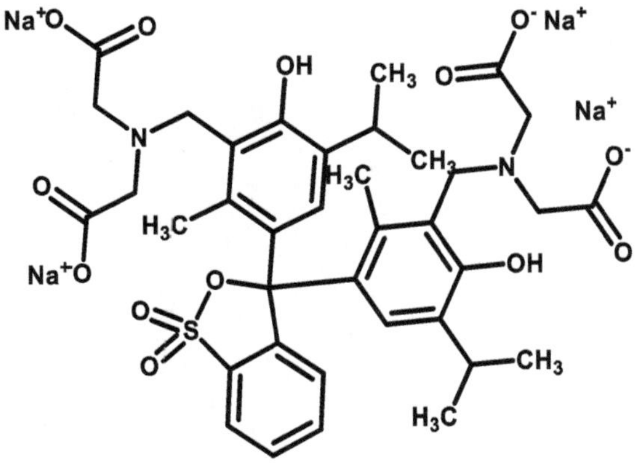

Figure 2.15: Xylenol orange.

2.4.3 Xanthene reagents

While numerous xanthenes reagents are available, the most commonly used are
1) gallelin
2) bromopyrogallol red
3) pyrogallol red.

They are used for spectrophotometric determination of Ti and Zr as well as Bi, In, Mo, Sn, Sb, Th, among others. In case a cationic surfactant is used, the sensitivity of such determination is enhanced. The structures of these three reagents are representing in Figs. 2.16, 2.17 and 2.18.

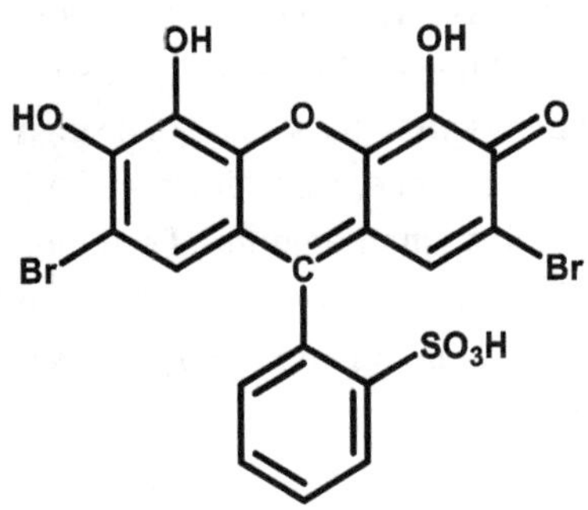

Figure 2.16: Bromopyrogallol Red.

Calcein also called fluorexone, fluorescein complexone, a reagent commonly used for calcium detection. It shares a similar structure, as illustrated in Fig. 2.18.

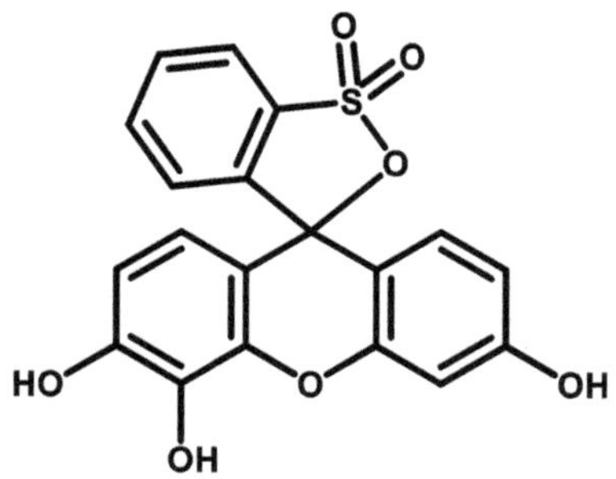

Figure 2.17: Pyrogallol Red.

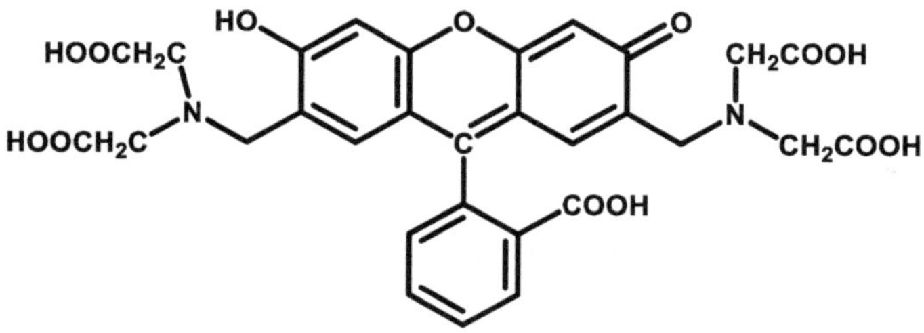

Figure 2.18: Calcein.

2.4.4 Non-chelating organic reagents

Basic dyes

These dyes are commonly used as halide salts. A number of elements form anionic complexes with such reagents. They many form ion pairs with basic dyes that are extractable into nonpolar solvents. Thus, these extracts are very useful for sensitive extractive spectrophotometric methods [26, 27]. The form of dye applied is an ion-associate of a singly charged cation. Solvents that are used include benzene, toluene, chloroform and 1,2-dichloroform ethane. To improve the extraction method, a small amount of a donor-active oxygen containing solvents like MiBK, butanol or even di-isopropyl ether may also be used. A moderately concentration mineral acid such as HCl and H_2SO_4 are used in the pH range 3–5 to control acidity of aqueous phase.

Triaryl methane dyes

Commonly used dye is triphenyl methane, which is included in the class of TAM dyes. The group is represented by brilliant green and malachite green. Their structures are represented in Figs. 2.19 and 2.20.

The other dyes included in this class are rhodamine 3B, butyl rhodamine B and rhodamine 6G, which are xanthene basic dyes.

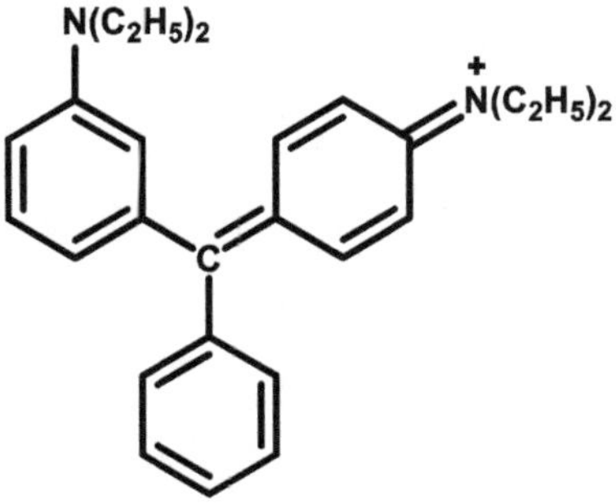

Figure 2.19: Brilliant green.

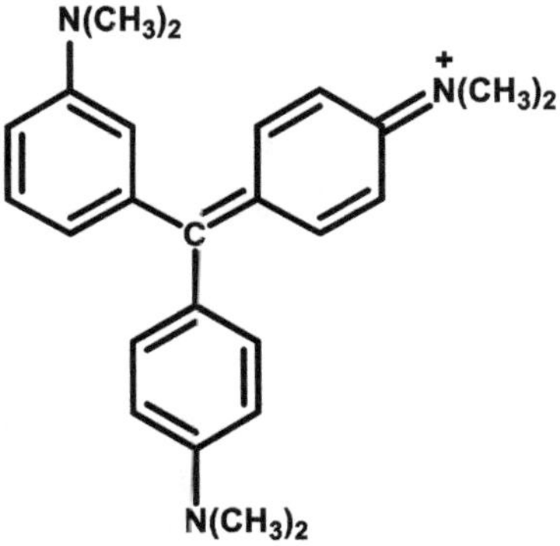

Figure 2.20: Malachite green.

2.4.5 Acid dyes

In general acid dyes (Figures 2.21–2.23) are less commonly used compared to basic dyes. They form extractable ion-associates with hydrophobic cationic metal complexes. They are commonly known as acid-base indications. The acid xanthenes dyes include erythrosine, rose bengal B and eosin. They are applied as sodium salts to facilitate the solubility; the following structures are representative of this category.

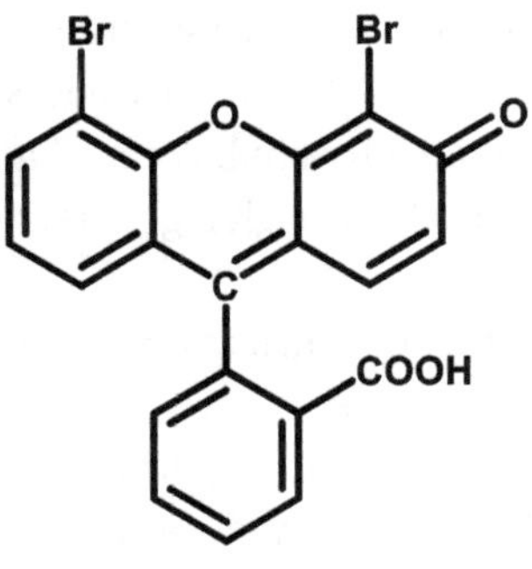

Figure 2.21: Eosin.

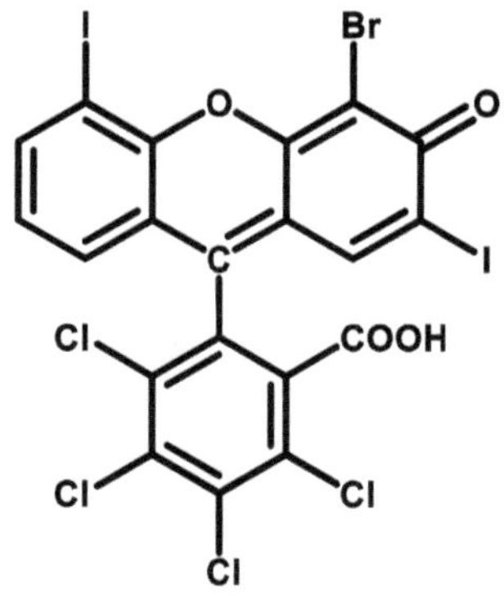

Figure 2.22: Erythrosin/rose bengal B.

Another dye which is of this class is bromophenol blue.

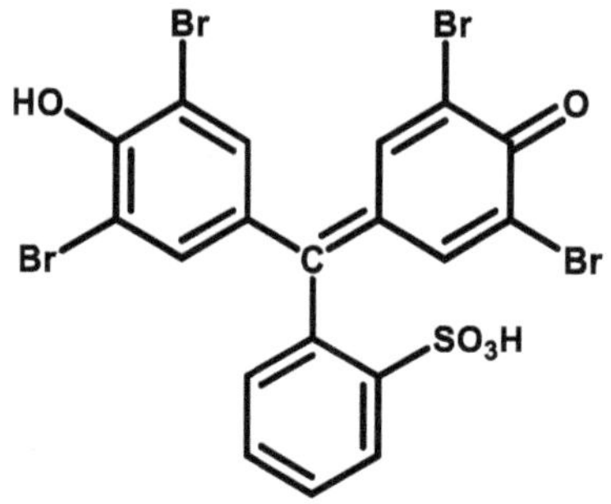

Figure 2.23: Bromophenol blue.

2.4.6 Dithizone

The most common and frequently used reagent is diphenylthiocarbazone, 3-mercapto-1,5-diphenyl-formazane, also called dithizone (H_2DZ). It is used in extractive separation and spectrophotometric determination of a number of metals [28–30]. The reagent is insoluble in aqueous solutions at pHs lower than 7, though it dissolves in basic media. Moreover, it dissolves in CCl_4, $CHCl_3$, hydrocarbons as well as alcohols giving a green solution. The structure is shown in Fig. 2.24.

The reagent binds metal through sulfur atom and other side is through nitrogen atoms [31].

Another analogue of this reagent is Di-(2-naphthyl) thiocarbazone (dinaphthizone), which is also a reagent for spectrophotometry, although the metal complexes of this reagent are less soluble compared to dithizone complexes.

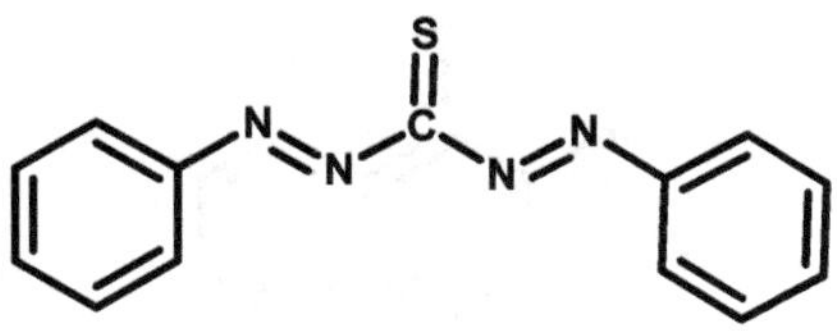

Figure 2.24: Dithizone.

2.4.7 Dithiocarbamates

A common reagent of this category is sodium diethyldithiocarbamate (Na-DDTC), Figure 2.25 for spectrophotometry [32–34].

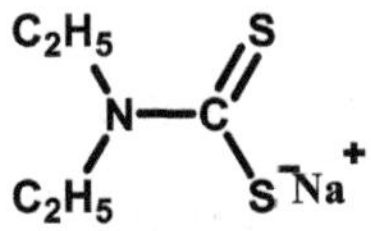

Figure 2.25: Dithiocarbamate.

The metal complexes of dithiocarbamates a poorly soluble in water. They dissolve in organic solvents like $CHCl_3$, CCl_4, acetone, etc. Figure 2.26 represents the structure of diethylammoniumdiethyldithiocarbamate,

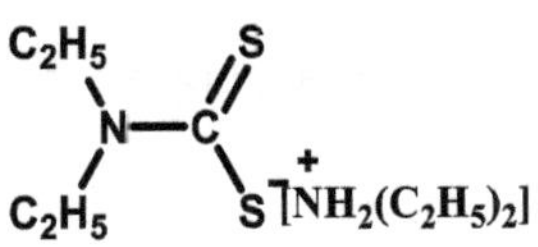

Figure 2.26: Dithiocarbamate.

The class of analytical reagent similar to DDTC diethylammonium is stable in acid solution but dissolves in $CHCl_3$. Common analogues of this category are dibenzyldithiocarbamate, pyrazolinedithiocarbamate and pyrolidinedithiocarbamate [35].

2.4.8 8-Hydroxyquinoline (Oxine)

The reagent 8-hydroxyquinoline (I), also called oxine, is a very common spectrophotometric reagent. The nature of the reagent is amphoteric, and it is soluble in $CHCl_3$, C_6H_6 and CCl_4, many other organic solvents. The distribution between solvent and water depends on the pH matrix of the solution. Other derivatives of oxine are chloro-oxine (II) and bromo-oxine (III). The structures are depicted in Figure 2.27.

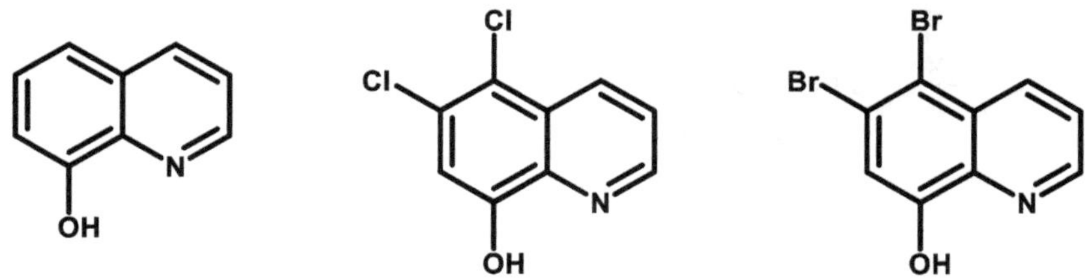

Figure 2.27: Oxines.

Bibliography

[1] Feigle, F., Anger, V., Spot Tests in Inorganic Analysis. 6, Elsevier, Amsterdam, 1972. Google Scholar.

[2] Fiegle, F., Quantitative analysis by spot tests: inorganic and organic applications. 2nd English edn, Elsevier, Amsterdam, 1958.

[3] Fiegle, F., Chemistry of Specific, Selective and Sensitive Reactions. Academic, New York, 1949 [Pub Med] [Google Scholar].

[4] Babko, A. K., Pilipenko, A. T., Fotometricheskii Analia (Photometric Analysis), Izd. Khimiya, Moscow, 1968.

[5] Alimarin, I. P. and Frid, B. I., Zavodskaya Laboratoria 6 (1937), 823.

[6] Alimarin, I. P., Medvedeva, Zavodskaya Laboratoria 11 (1945), 254.

[7] Sandell, E. B., Onishi, H., Photometric Determination of Traces of Metals, General Aspects (Chapters 5 and 6), Wiley, New York, 1978.

[8] Cheng, K. L., Uenok, Imamura, T., Handbook of Organic Analytical Reagents, CRS Press, Boca Raton, 1982.

[9] Sommer, L., Ackermann, G., Burins, D. T., Savvin, S. B., Pure and Applied Chemistry 62, (1990), 2147, 232.

[10] Townshend, A., Burns, D. T., Guibault, G. G., Lobinski, R., Marczenkoz, N. E. J., Onishi, H., Dictionary of Analytical Reagents, Chapman and Hall, London, 1993.

[11] Stephen, W. L., Analyst 102 (1977), 793.

[12] Nawaz-Diaz, A., Talanta 92 (1991), 207.

[13] Dalawat, D. S., Chauhan, R. S., Goswami, A. K., Review of Spectrophotometric methods for determination of zirconium, Reviews in Analytical Chemistry 24(2) (2005), 75–102.

[14] Goswami, A. K., Agarwal, S., Spectrophotometric Determination of Copper and Iron Methods and Reagents, DeGruyter GMBH, Berlin/Boston, 2021.

[15] Goswami, A. K., Spectrophotometric Determination of Nickel and Cobalt Methods and Reagents, De-Gruyter GMBH, Berlin/Boston, 2024.

[16] Goswami, A. K., Spectrophotometric Determination of Palladium and Platinum Methods and Reagents, CRC Press, Boca Raton F. (USA), 2023.

[17] Goswami, A. K., Spectrophotometric Determination of Vanadium, Chromium and Manganese Methods and Reagents, DeGruyter GMBH, Berlin/Boston, 2024.

[18] Anderson, R. G. and Nickless, G., Analyst 92 (1967), 207.

[19] Pilipenko, A. T. and Sarvansky, L. I., Talanta 25 (1978), 207.

[20] Mapuliya, S. G., et al., Zhurnal Analiticheskoi Khimii 35 (1980), 1307.

[21] Hovind, H. R., Analyst 100 (1975), 769.

[22] Savvin, S. B., Organic Reagents of Arsenazo III group, Atomizdat, Moscow, 1971.

[23] Ivanov, V. M. and Rudometkina, T. F., Zhurnal Analiticheskoi Khimii 33 (1978), 2446.

[24] Savos'kina, L. N., Nemodrunk, A. A., Oskotskaya, E. P., Zhurnal Analiticheskoi Khimii 36 (1981), 1483.

[25] Pringle, D. L., et al., Talanta 29 (1982), 197.
[26] Zadorozhnaya, E. M., Nabivanets, B. I., Masley, N. N., Zhurnal Analiticheskoi Khimii 29 (1974), 993.
[27] Marczenko, Z., Mikrochimica Acta II (1977), 651.
[28] Iwantscheff, G., Das Dithizone Lind Seme Auwendung in der Mikro-Lind Spine analyze, 2nd ed., Verlag Chemie, Weinheim, 1972.
[29] Irving, N. M., Dithizone, Royal Society of Chemistry, London, 1977.
[30] Irving, N. M., Critical Reviews in Analytical Chemistry 8 (1980), 321.
[31] Lang, M. and Alsop, P. A., Talanta 17 (1970), 242.
[32] Hulanicki, A., Talanta 14 (1967), 1371.
[33] Halls, D. J., Mikrochimica Acta 62 (1969).
[34] Bayo, S., Analytica Chimica Acta 105 (1979), 281.
[35] Likussar, W., Boltz, D. F., Analytical Chemistry 43 (1967), 1273.

3 Spectrophotometric determination methods for lanthanides

In the preceding section, methods using various analytical reagents were described. Since lanthanum and other lanthanides have been reported to have similar or common methods, their determination has been compiled under the common term of lanthanides. An attempt has been made to include as far as possible most relevant details like pH, λmax, solvent, interference and molar absorptivities, yet there may be some data missing. However, the details given are user friendly and can be applied directly by any analyst.

3.1 2,3-Dihydroxynaphthalene

Tarafdar, Pradhan et al. [1] developed a liquid–liquid extraction based spectrophotometric method for the determination of rare earths using 2,3-dihydroxynaphthalene as an extractant cum reagent, for geological samples. Lanthanides form 1:3 complexes with 2,3-dihydroxynaphthalene and are extracted in ethyl acetate from aqueous solution at pH > 8. At pH 6.3, the lanthanides are stripped off and identified using arsenazo III.

3.2 Arsenazo III

Babula et al. [2] presented a rapid spectrophotometric method for the determination of lanthanum (III) ions in water samples. The article also includes optimization studies on spectrophotometric determination of lanthanum (III) in water samples.

3.3 Morin

Kopacz et al. [3] proposed a method for spectrophotometric determination of lanthanides La (III), Sm (III), Tb (III), Dy (III), Ho (III) and Er (III) ions, using morin as a complexing reagent. The solid complexes formed were confirmed in terms of the molecular formula Ln $(C_{15}H_9O_7)_3.nH_2O$, where $n = 6$ for La, Sm, Tb, Dy and $n = 5$ for Ho and Er.

3.4 Diazepam

Khalile et al. [4] described a simple accurate method for determination of Pr^{3+}, Nd^{3+} and Sm^{3+} using diazepam as a reagent. The spectral measurements were reported at 366 nm in the concentration range of 10–140 μg/mL. The Sandell's sensitivity is reported

https://doi.org/10.1515/9783111615004-003

as 160×10^{-3} to 3.8×10^{4} in the optimum concentration of 36.9–125.8 µg/mL. A general method for the determination of rare earth metals using diazepam is thus successfully developed.

3.5 Polyvinyl alcohol condensed p-formylchlorophosphanazo

Jin et al. [5] proposed a combined determination method using a solid–liquid extractive preconcentration with polyvinyl alcohol condensed p-formylchlorophosphanazo as the reagent. This is claimed to be a highly sensitive and inexpensive method for determining rare earth elements with cloud point extraction before spectrophotometric determination. The measurements were made at 668 nm in the Beer's Law range for a mixture of $RE_{(x)}O_{(y)}$–PV–CPAPF 0.1–2 mg/L, and the molar absorptivity value is 7.26×10^{4} mol/cm/l. The procedure has been applied to the total determination of rare earth metals in ductile iron and micronutrient fertilizers, with satisfactory results.

3.6 2-mercapto-4,4,6-trimethyl – (1H,4H)-pyrimidine-1-acetic acid

Kumar Ashok [6] has proposed a method for spectrophotometric determination of lanthanides *viz.*, La (III), Pr (III), Nd (III), Sm (III), Gd (III) and Tb (III) using extraction with 2-mercapto-4,4,6-trimethyl-(1H,4H)–pyrimidine-1-acetic acid as an extractant from dilute aqueous nitrate media; the reagent can extract mentioned ions that are subsequently determined spectrophotometrically. It is mentioned that maximum enhancement is achieved in presence of 40 % acetonitrite, and very poor extractability in some cases is supported by using calcium nitrate as a salting-out agent.

3.7 Simultaneous determination model

Peralta-Zamora et al. [7] proposed a chemometric alternative for resolving the classical spectrophotometric determination of lanthanide mixtures, which is the greatest problem in analytical chemistry owing to the close similarity of properties. A model based on multivariate calibration (Partial Least Squares Regression, PLSR) was proposed. It is reported that the model uses absorption spectra in the range 290–800 nm for a set of 20 mixtures of Ce, Pr, Nd and Sm and the determination of Ce, Pr and Nd in concentrations of a commercial rare-earth product is possible. Determination of Sm is not possible using the model, as mentioned.

3.8 N-phenylcinnamohydroxamic acid

Agarwal and Thomaskutty, [8] developed a solvent extraction and spectrophotometric method for the determination of lanthanides using N-phenylcinnamohydroxamic acid

(PCHA) as an extractant. This is described that the reagent forms a 1:3 yellow-colored binary complex with La (III), Pr (III), Nd (III), Sm (III), Gd (III) and Tb (III) which are extractable with chloroform at pH 9.0. These complexes absorb at 380 nm, and if xylenol orange is added, the extract has maximum absorbance between 590–610 nm. Further, the possibility of separation of lanthanides from each other has also been studied. The method has been applied and validated by determination in standard United State Geological Samples (USGS) and IAEA soil sample, too.

3.9 Arsenazo III

Botella et al. [9] proposed a sensitive method without separation for total lanthanides using arsenazo III in stainless-steel samples. It is reported that arsenazo III forms a pinkish-crimson complex with lanthanides at pH 1 and absorbs at 650 nm. In the presence of 60–70 % alcohol, the Beer's law in the range of 1–40 µg/50 mL is applicable, and selectively lanthanides can be determined in stainless-steel samples.

3.10 1-(2-pyridylazo)-2-naphthol (PAN)

Mathew et al. [10] have developed a spectrophotometric method for the determination of Nd (III), Sm (III), Gd (III), Tb (III), Dy (III) and Ho (III) in a micellar media using PAN as a chromogenic reagent. This method is based on the formation of an intenselycolored complex ML_3 in the pH range of 8.5–9.5, and the micellar media solubilize the complexes. The commonly used micellar media are sodium dodecyl sulfate, cetyltrimethyl ammonium bromide and Triton X–100. The reported molar absorptivity values for each medium are $6.2–9.8 \times 10^4$, $4.4–8.5 \times 10^4$ and $4.3–7.8 \times 10^4$ mol/cm/dm^3, whereas Sandell's sensitivity for Triton X–100, CTAB and SDS are $1.5–3.02\times10^{-4}$, $1.7–2.96\times10^{-4}$ and $2.1–2.8\times 10^{-4}$ µg/cm^2. Summary of analytical reagents and their parameters are depicted in Table 3.1.

3.11 Norfloxacin

Wang et al. [11] have developed a method for determination of neodymium, holmium and erbium in mixed rare earths using norfloxacin as a reagent. It is described that norfloxacin in the presence of cetylpyndinum chloride forms a complex with the metals. The reported wavelengths are 57 nm for neodymium, 450 nm for holmium and 523 nm for erbium.

The Beer's law range is $5.0 \times 10^{-5}–2.5 \times 10^{-4}$ mol/L. This method can determine mixtures of rare earths including Nd (III), Ho (III) and Er (III) successfully.

Table 3.1: List of analytical reagents for determination of lanthanides.

S. No.	Name of Reagent	λ_{max} pH/Solvent	Molar Absorptivity L/mol/cm	Sandell's Sensitivity μg/cm^2
1.	2,3-Dihydroxy-naphthalene	>8.0	-	-
3.	Morin	-	-	-
4.	Diazepam	366	-	160×10^3 to 3.8×10^4
5.	Polyvinyl alcohol condensed p-formylchlorophosphazane	668	7.26×10^4	-
6.	2-mercapto-4,4,6-trimethyl-(1H,4H)-pynmidine-1-acetic acid	-		
7.	Chemometric method	-		
8.	N-phenyl-cinnamohydroxamic acid +xylenol orange	590–610		
9.	Arsenazo III	650 pH 1	-	-
10.	1-(2-Pyndylazo)-2-naphthol (PAN)	8.5–9.5	$6.2–9.8 \times 10^4$ $4.4–8.5 \times 10^4$ $4.3–7.8 \times 10^4$	$1.50–3.02 \times 10^{-4}$ $1.7–2.96 \times 10^{-4}$ $2.1–2.8 \times 10^{-4}$
11.	Norfloxacin	575 450 523	-	-

Bibliography

[1] Tarafdar, P., Mondal, S., Liquid-liquid extraction of rare earth elements (total) using 2,3-dihydroxy naphthalene as an extractant and its application to the determination of REEs (Total) in rock, soil and sediment samples by spectrophotometry, Journal of the Indian Chemical Society 90(11) (2013), 1961–1967.

[2] Babula, P., Beklova, M., Svobodova, Z., Majzlik, P., Adam, V., Kizek, R., Rapid spectrophotometric determination of lanthanum (III) ions in water samples, Listy Cukrovarnicke a Reparske 126(II) (2010), 378–398.

[3] Kopacz, W., Complexes of Ca (III), Sm (III), Tb (III), Dy (III), Ho (III) and Er (III) ions with morin, Polish Journal of Chemistry 78(4) (2004), 521–528.

[4] Khalile, Z., Abd Allah, Spectrophotometric determination of diazepam in pure form and in pharmaceutical preparation, Egyptian Journal of Chemistry 47(1) (2004), 1–16.

[5] Jin, G., Kan, J., Zhu, Y., Lei, V., Chen, B., Spectrophotometric determination of the total amount of rare-earth after solid-liquid extractive preconcentration with polyvinyl alcohol condensed p-formylchlorophosphonazo, Microchemical Journal 64(1) (2000), 111–118.

[6] Kumar, A., Liquid-liquid extraction of trivalent lanthanides from nitric acid by 2-mercapto-4, 4, 6-trimethyl-(1H, 4H)-pyrimidine-1-acetic acid, Indian Journal of Chemical Technology 4 (1997), 57–58.

[7] Peralta-Zamora, P., Cornejo-Ponce, L., Nagata, N., Poppi, R. J., Chemometric alternatives for resolution of classical analytical problems; Spectrophotometric determination of lanthanide mixtures, Talanta 44(10) (1997), 1815–1822.

[8] Agarwal, Y. K., Thomaskutty, P. T., Solvent extraction and spectrophotometric determination of lanthanides with N-phenyl cinnamohydroxamic acid, Indian Journal of Chemistry, Section A: Inorganic, Physical, Theoretical and Analytical 32(1–3) (1993), 277–280.

[9] Botella, B. A., Colorimetric determination of lanthanides in stainless steels, Revista de Metalurgia 24(2) (1988), 271–275.

[10] Mathew, A., Kumar, K., Shyamala, Satyanarayana, Rao, Spectrophotometric determination of Neodymium (III), Samarium (III), Gadolinium (III), Terbium (III), Dysprosium (III), Holmium (III) in micellar media, Indian Journal of Chemical Technology 19(5) (2012), 331–336.

[11] Wang, N.-X., Wang, L., Jiang, W., Ren, Y.-Z., Si, Z.-K., Qiu, X.-X., Du, G.-Y., Qi, P., Determination of neodymium holmium and erbium in mixed rare earths by norfloxacin, Fresenius, Journal of Analytical Chemistry 361(8) (1998), 821–824.

4 Spectrophotometric methods for determination of (A) lanthanum (B) scandium

4.1 Methods for spectrophotometric determination of lanthanum

4.1.1 Bromopyrogallol red and cetylpyridinium chloride

Saber and Mohammad [1] proposed a spectrophotometric determination method for La (III) using bromopyrogallol red and cetylpyridinium chloride in natural water and urine. The method is claimed to be highly sensitive, and it is explained that La (III)–BPR–CPC complex is blue in color, and the ternary complex absorbs at 661 nm. A 0.125–5.0 µg/mL range is linear and applicable for the determination following Beer's law. The reported value of molar absorptivity is 5.39×10^4 mol/cm, and Sandell's sensitivity is 0.00258 µg/cm^2 for the binary complex. The limit of detection is 0.0630 µg/mL, and the quantitation limit is 0.0817 µg/mL. The method is successful in the determination of La (III) in natural water and urine samples.

4.1.2 Tribromoarsenazo

Chen et al. [2] proposed a spectrophotometric determination method for lanthanum (III) using tribromo arsenazo as a reagent. It was reported that addition of ascorbic acid masks any interference of other interfering species. The optimal experimental conditions consisted of 10 g/L ascorbic acid (5 mL), 0.04 mol/L HCl, and 0.5 g/L tribromoarsenazo (2 mL). Beer's law is obeyed in the range 0–10 µg (25 mL^{-1}) for lanthanum, with an extraction coefficient > 0.99. The wavelength range of determination is 580–560 nm. The method is successful in the determination La (III) as mentioned.

4.1.3 Crown hydroxamic acid

Agrawal and Shrivastava [3] developed a new spectrophotometric method for the determination of lanthanum (III) using crown hydroxamic acid as a chromogenic reagent. It is documented that a new crown hydroxamic acid 5,14-N,N'-hydroxyphenyl-4,15-dioxo-1,5,14,18-tetraaza hexacosane (NHDTAHA) forms a yellow-colored complex with La (III), which is extracted in chloroform. This complex absorbs at 372 nm, in Beer's law range of 1.2–20 ppm, and has molar absorptivity of 7.7×10^3 l/mol/cm. The detection limit of 5.140 ppb of lanthanum (III) is achieved using this method.

https://doi.org/10.1515/9783111615004-004

4.2 Methods for spectrophotometric determination of scandium

4.2.1 Introduction

The element scandium, with atomic number 21, represented by symbol Sc, is found in very small amounts in nature. Since it is mined from rare earth minerals from Scandinavia, it is given the name scandium. The abundance of scandium ranks as 50th among the elements on Earth; it is distributed in more than 800 minerals with a total production at about 50 kg/year. The metal is obtained by heating scandium fluoride with calcium to give pure metal. The element has applications in many devices and glasses such as lamps and televisions, among others. During welding of high strength aluminum alloys, it is used to reduce solidification and cracking. Aluminum–scandium alloys are used in the aerospace industry and even sports equipment such as baseball and cycling, to name a couple. When exposed to air, it changes to yellowish or pinkish color. It dissolves in acids and reacts with hydrogen gas. The metal belongs to the transition metal group, and due to its limited availability and high cost, it has limited applications. With a high melting point and low density, it's used as an alloying agent to improve performance in military applications. On the downside, however, the metal concentrates and can harm the liver. Although the metal itself is not toxic, some of its compounds are carcinogenic, and although the occurrence is rare, it has been found to enter into the food chain in trace amounts. Even if a accumulates less than 0.1 microgram per day, it is dangerous especially due to the fact that the gases and damps impregnated with scandium can be inhaled in air. It damages cell membranes of aquatic animals causing negative impact on their reproduction and nervous system. Since both aspects, negative and positive, warrant its determination in a variety of samples, the next section describes methods for its spectrophotometric determination.

4.2.2 Various spectrophotometric methods for the determination of scandium

This section briefly describes the method's details (pH, molar absorptivity, Sandell's sensitivity etc.) to inform readers of the available a complete procedure. The methods are presented in chronological order so that oldest to the latest methods can be viewed easily at the same time. The method list includes title of the reagent used. An attempt has been made to include all the methods in tabular form in Table 4.1 in summary form.

4.2.3 Arsenazo compounds

Onishi and Banks [4] used arsenazo for the spectrophotometric determination of scandium in the range of 10–50 µg. The solution with pH 6.1 has an absorbance at 570 nm. It is shown that scandium is separated from magnesium, calcium, other rare earths,

Table 4.1: Common analytical reagents used for spectrophotometric determination of scandium.

	Name of the reagent	λ_{max} nm pH/Solvent	Molar absorptivity L/mol/cm	Sandell's sensitivity ng/cm^2	Ref.
1.	Arsenazo	570 6.1	–	–	
2.	Chromazurols	550	27000	.0017	
3.	Tiron	310 6.0	8×10^3	–	
4.	Bromopyrogallol Red	610	2.4×10^4	–	
5.	Anthrarufin-2,6-disulphonic acid (disodium salt)	–	–	–	
6.	Nalidixic acid	640 4.5	–	–	
7.	p-Acetylchlorophosphanazo		2.29×10^5	–	
8.	o-Hydroxy hydroquinonaphthalein	– 4.0	1.0×10^5	0.0005	
9.	Semi xylenol orange	530	2.4×10^4	–	
10.	Indoferron	800	960	–	
11.	Eriochromcyanin-R		1.8×10^5	–	
12.	XAD-4 resin impregnated with nalidixic acid	640 4.5	–	1.4 µg/L 0.32 µg/L	
13.	Chlorosulphophenol C	649 2.0	7.78×10^5	–	
14.	Hexaacetate calyx (6) arene	675 1 M Nitric acid (5 pH)	–		
15.	Eriochrome cyanon R (ECR)	610 6.5/ (1.0 M HCl)	5.6×10^5 1.0×10^7 M (DL)		
16.	Nile blue and poly (vinyl alcohol)	585	3.59×10^5	–	
17.	2-(5-Bromo-2-pyridylazo)-5-diethyl-aminophenol	572 5.3 – 6.5	7.2×10^4	0.62	
18.	N-phenylbenzo-hydroxamic acid (PBHA)	540 4.5	1.7×10^4		
19.	o-Hydroxyhydroquinohaphthalein in cationic micellar media	5.4	1.0×10^5	0.0005 µg/cm^2	
20.	Pyridylazo and (Thiazolylazo) naphthols	545 570	1.9×10^4 to 2.93×10^4		

zirconium, fluoride and phosphate by extraction with TTA in xylene. Furthermore, copper, aluminum and iron (iii) are also extracted with 8-quinolinol-chloroform extraction. Even uranium (vi) and thorium can be separated using anion exchange. The method successfully determines scandium even in complex mixtures, as claimed.

4.2.4 Chromazurol S

Ryoei and Norio [5] developed a spectrophotometric method for the determination of scandium using chromazurol-S in a weakly acidic medium. The λ_{max} of the complex is 550 in the concentration range between 4.27 µg/10 mol. The pH for the best color development is 5.6. The molar extinction coefficient reported is 27,000 with a Sandell's sensitivity of 0.0017 µg/cm^2. It is further reported that many ions, such as fluoride, phosphate, oxalate, tartaric citrate and EDTA, interfere. Moreover, cations like Al (iii), Be (ii) U (vi), Fe (ii), Th (iv), Cu (ii), Y (iii), Fe (iii) and Zr (iv), as well as W (vi), interfere in the determination as mentioned by the authors.

4.2.5 Tiron

Hamaguchi et al. [6] reported a method for the spectrophotometric determination of scandium with Tiron, and it is also documented that 3–250 µg of scandium can be determined with Tiron. A 1:3 and 1:1 complex of the scandium with Tiron is suggested, with the pH 6.0 in acetate buffer. The reported value of extinction coefficient is 8×10^3 at 310 nm. The authors report interference of small amounts of aluminum, titanium, iron, cerium molybdenum, thorium and zirconium. Furthermore, the interference of oxalate, citrate, tartrate and malate is also reported.

4.2.6 Bromopyrogallol red

Shimizu [7] reported a method for the spectrophotometric determination of scandium using bromopyrogallol red. This method as highly sensitive and can detect a microgram of scandium. The report mentions that at 610 nm and pH between 6 and 1, scandium can be determined successfully. A 1:1 complex at 610, with molar absorptivity of 2.4×10^4, is reported. Although common cations do interfere, this yields complete separation by successive ion-exchange. It also mentions that scandium in traces can be determined with several igneous rocks.

4.2.7 Anthrarufin-2,6-disulforic acid (disodium salt)

Macdonald and Yoe [8] developed a method for the spectrophotometric determination of scandium using anthrarufin-2,6-disulforic acid (disodium salt) as an analytical reagent; up to 2-ppm of scandium can be determined using this method. It is mentioned that 0.04 µg of scandium can be determined due to the sensitivity of this method. A separation procedure has also been developed to remove interference.

4.2.8 Nalidixic acid

Shahida and Khan [9] reported an online preconcentration-based spectrophotometric method for the determination of scandium using naldixic acid as an analytical reagent. The method describes that scandium, from an aqueous sample solution of pH 4.5, it's selectivity retained in a mini column containing XAD-4 resin impregnated with nalidixic acid, with a flow rate of 1.18 mL/min, in the form of a scandium naldixic acid complex. This complex is desorbed using an arsenazo III solution and finally measured at 640 nm. The detection limit of the scandium is 1.4–0.32 μ/L. It is mentioned that tolerances from cations such as Th (iv), U (vi) and some rare earths, and some anions have been improvised. The method as mentioned has been validated on certified materials IAEA-SL-1 (lake sediment), and results for scandium are in good agreement.

4.2.9 p-Acetylchlorophosphanazo (CPAPA)

Hsu et al. [10] developed a spectrophotometric method the for determination of scandium based on complex formation of a scandium-cerium-p-acetylchlorophosphanazo system, and tit is mentioned that a heteronuclear ternary β-complex is formed. The scandium possesses a molar absorptivity $\varepsilon = 2.29\times102$ l/mol/cm even in the presence of many interfering ions. The method has been extended to scandium in alloys with good results.

4.2.10 o-Hydroxyhydroquinonephthalein, cationic micellar media

Mori et al. [11] reported a spectrophotometric method for the determination of scandium (iii) by using ternary complex with o-hydroxyhydroquinonephthalein (Qnph) and N-hexadecyl-pyridinium chloride (HPC) at pH 4.0 as a cationic surfactant. The determination occurs at 555 nm, with molar absorptivity of 1.0×10^5 dm^3 mol^{-1} cm^{-1} and Sandell's sensitivity of 0.0005 μg/cm^2.

4.2.11 Semi-xylenol orange

Hafez et al. [12] developed a method for spectrophotometric determination of scandium (iii) using semi-xylenol orange. In a medium of chloroacetate buffer, the complex scandium-semi xylenol orange is formed and measured at 530 nm. The reported molar absorptivity is 2.4×10^4 L/mole/cm.

4.2.12 Indoferron

Shimizu and Ogami [13] introduced a new spectrophotometric method for microgram amounts of scandium using indoferron as a reagent. It is measured at 600 nm and has a molar absorptivity of 960 L/mol/mm. The method is free from interference of uranium and other rare earths. It is noted that scandium is separated by a three-stage ion-exchange chromatographic procedure.

4.2.13 Eriochromecyanin

Syndorova et al. [14] reported a spectrophotometric determination method for Sc (iii) using eriochromecyanin as a reagent. The effect of cationic water-soluble polyelectrolyte PSPMG on the complex was studied. The method involves formation of the complex at pH 5.5–6.5 when scandium, ECR and PSPMG in the buffer solution are added, resulting in a 1:4:1 complex. The reported molar absorptivity value is 1.8×10^5, and the limit of detection is 10^{-7} mol/L of Sc.

4.2.14 Chlorosulphophenol C

Deng et al. [15] developed a spectrophotometric method for scandium determination in an HCl leaching solution of red mud. They mentioned that the reagent chlorosulphophenol C and Sc (iii) form a 1:1 complex at pH 2.0 at 649 nm, which can be measured. The reported molar absorptivity value is 7.78×10^5 L/mol/cm, and the method can be applied to the determination of scandium (iii) in HCl acid-leaching solution of red mud.

4.2.15 Hexaacetato calix (6) arene

Swamidoss et al. [16] described a method for extraction based on the spectrophotometric determination of scandium (iii) with hexacetate calyx (6) arene. They detailed that the solvent extraction of Sc (iii) was carried out with hexacetate calyx (6) arene in toluene, and Sc (iii) was quantitatively extracted with 1×10^{-4} M of ligand at pH 5 in toluene. The time of equilibration was 12 minutes. Subsequently, the organic phase was stripped with 1M of nitric acid to determine Sc (iii) at 675 with arsenazo III. The report also states that they successfully extended the method in real samples.

4.2.16 Amberlite IRC 718 resin

Park and Cha [17] proposed a spectrophotometric method for scandium with eriochrome cyanin R (ECR). At pH 6.5, 5 ml of 1.0×10^{-2} M ECR, 10 ml of 2.0×10^{-4} M, CTMABA form a ternary complex with Sc (iii). This can be measured at 610 nm with a detection limit of 1.0×10^{-7} M. The method was applied to the determination of Sc (iii) in monazite.

4.2.17 Molybdate and Nile Blue

Li et al. [18] developed a sensitive spectrophotometric method for the determination of trace amounts of scandium using spectrophotometry. The method is based on the formation of ion associate between scandium (iii), molybdate and Nile Blue in the presence of polyvinyl alcohol. The measurement was done at 585 nm, and the reported molar absorptivity was $3.59 \times 10^5\ dm^3\ mol^{-1}\ cm^{-1}$. Beer's law linear range was 0–2.0 µg scandium/25 ml. The advantage reported for the method is that tenfold amounts of rare earths do not interfere the determination except cerium. The method was successful even for geological samples for measuring scandium (iii).

4.2.18 2-(5-Bromo-2-pyridylazo)-5-diethylaminophenol

Rathaiah and Eshwar [19] published a method for scandium (iii) with 2-(5-bromo-2-pyridylazo)-5-dithylaminophenol as a chromogenic reagent. A 1:1 red-colored complex in the pH range 5.3–6.5 was reported; it is water insoluble but dissolves in > 1 = 32 % (v/v) methanol. It was measured at 572 nm. The reported molar absorptivity was 7.3 E4 $dm^3\ mol^-\ cm^-$, and the Sandell's sensitivity value was 0.62 ng/cm^2. Additionally, a coefficient of variation of +/−0.43 % with no interference of 50 ions was mentioned. The method as described can be used for the determination of Sc (iii) in wolframite.

4.2.19 N-Phenylbenzohydroxamic acid

Agrawal and Nagar [20] proposed a method of scandium determination using N-phenylbenzo-hydroxamic acid as a reagent. In the article, they described how scandium (iii) is extracted with N-phenylbenzohydroxamic acid into isoamylalcohol at pH 4.5. A further mixed complex is formed with xylenol orange (XO) or morin. The SC-PBHA–XO/morin Mn complex, a red-colored one, is measured at 540 nm in the Beer's law range of 0.05–3.0 ppm. It has a molar absorptivity of $1.7\ 10^4\ dm^3\ mol^{-1}\ cm^{-1}$ at 540 nm. It has also a greenish yellow fluorescence so that scandium can be determined fluorometrically too in the range of 0–1.4 ppm as mentioned.

4.2.20 (Pyridylazo)-and (thiazoylazo) naphthols

Sharma et al. [21] developed a rapid spectrophotometric method for the determination of scandium (iii) with the reagent 1-(2-pyridylazo)-2-naphthol (PAN) and 1-(2-thiazoylazo)-2-naphthol (TAN) in weakly acidic media. The complexes formed are 1:1 and 1:2 chelates with the reagents. The PAN and TAN complexes are reportedly determined at 545 and 570 nm in the Beer's law ranges of 0.16–1.40 and 0.08–1.40 scandium (iii), respectively.

The molar absorptivity values are 1.90×10^4 and 2.93×10^4 dm^3 mol^{-1} cm^{-1}. The effect of diverse ions on the determination has also been reported.

Bibliography

[1] Saber, A. I., Mohammad, S. A., Spectrophotometric determination of lanthanum (III) with bromopyrogallol red and cetylpyndinium chloride. application in national waters and urine, Journal of Pharmaceutical Negative Results 13 (2022), 3117–3125.

[2] Chen, X., Nn, L., Kuang, L. F., Long, H., Spectrophotometric determination of lanthanum in raffinate with tribromoarsenazo, Applied Mechanics and Materials 675–677. (2014), 1429–1432.

[3] Agrawal, Y. K., Shrivastava, P., Solvent extraction, spectrophotometric and inductively coupled plasma emission spectroscopic (ICP-AES) determination of lanthanum (III) with crown hydroxamic acid, Talanta 44(7) (1977), 1307–1312.

[4] Onishi, H., Banks, C. V., Spectrophotometric determination of scandium with Arsenazo, Analytica Chimica Acta 29 (1963), 240–248.

[5] Ryoei, I., Norio, H., The spectrophotometric of scandium (iii) with chromazurol S, Bulletin of the Chemical Society of Japan 40(5) (1967), 1153–1158.

[6] Hamaguchi, H., Onuma, N., Kuroda, R., Sugisita, R., Ultra violet spectrophotometric determination of scandium with Tiron, Talenta 9(7) (1962), 563–571.

[7] Shimizu, T., Spectrophotometric determination of scandium with bromopyrogallol red, Talanta 14(4) (1967), 473–479.

[8] Macdonald, C., Yoe, J. H., Spectrophotometric determination of scandium with anthrarufin-2, 6-disulfonic acid (disodium salt), Analytica Chimica Acta 28 (1963), 264–270.

[9] Shahida, S., Ali, A. and Khan, M. H., Online spectrophotometric determination of scandium after preconcentration on XAD-4 resin impregnated with nalidixic acid, Journal of the Iranian Chemical Society 10(3) (2013), 461–470.

[10] Hsu, C.-G., Liu, S.-C., Pan, J.-M., Spectrophotometric determination of scandium based on the coloration effect in scandium-cerium-p-acetyl chlorophosphanazo (CPAPA) system, Talanta 42(12) (1995), 1905–1911.

[11] Mori, I., Fujita, Y., Fujita, K., Usami, A., Kawbe, H., Koshiyama, Y. and Tanaka, T., Spectrophotometric determination of scandium (iii) with o-hydroxyhydroquinonaphthalein in catioinic micellar, Media Bulletin of the Chemical Society of Japan 59(5) (1986), 1623–1625.

[12] Hafez, M. A. H., Kenaway, I. M. M., Kabil, M. A., Spectrophotometric investigation and determination of scandium using semi-xylenol orange, Analytical Letters 18(16) (1985), 2043–2067.

[13] Shimizu, T., Ogami, K., Analytical application and spectrophotometric study of the scandium indoferron complex, Talenta 16(12) (1969), 1527–1533.

[14] Sydorova, L., Chernyavska, A., Chmylenko, T., Chmyleuko, F., Polyelectrolytic effect on spectrophotometric determination and interaction of scandium (iii) with eriochromcyanin, Eastern-European Journal of Enterprise Technologies 4(6-88) (2017), 53–58.

[15] Deng, H., Wang, K., Wang, H., Li, A., Zhang, F., Chen, J., Spectrophotometric determination of scandium in hydrochloric acid leaching solution of red mud, Xiyou Jinshu/Chinese Journal of Rare Metals 35(5) (2011), 764–768.

[16] Swamidoss, Khopkar, Malkhede, Liquid extraction and spectrophotometric determination of scandium (iii) with hexacetato calix arene, Extraction and Research for Atomic Minerals 19 (2009), 211–218.

[17] Park, C.-I. and Cha, K.-W., Spectrophotometric determination of scandium-(iii) in monazite after separation using amberlite IRC 718 chelating resin, Bulletin of the Korean Chemical Society 20(12) (1999), 1409–1412.

[18] Li, Z., Wang, J., Xu, Q., Spectrophotometric determination of trace amounts of scandium with molybdate, ile blue and polyvinyl alcohol, Analytical Sciences 12(2) (1996), 259–262.
[19] Rathaiah, G. V., Eshwar, M. C., Spectrophotometric determination of Sc (iii) with 2-(5-bromo-2-pyridylazo)-5-diethylaminophenol Indian, Journal of Chemistry, Section A, Inorganic Physical, Theoretical and Analytical 25(1) (1986), 101–102.
[20] Agrawal, Y. K., Nagar, A. K., Solvent extraction and spectrophotometric determination of scandium (iii) with N-phenylbenzohydroxamic acid, Indian Journal of Chemistry Section A. Inorganic, Physical Theoretical Analytical 25(11) (1986), 1065–1066.
[21] Sharma, C. D., Nagarkar, Sujan, S. G. and Eshwar, M. C., Rapid spectrophotometric determination of scandium (iii) with (pyndylazo) – and (thiazolylazo) naphthols, Bulletin of the Chemical Society of Japan 59(5) (1986), 1662–1664.

5 Spectrophotometric methods for the determination of (A) cerium (B) miscellaneous rare earths

5.1 Spectrophotometric methods for the determination of cerium

5.1.1 2,6-Dibromo-4-chlorocarboxy arsenazo

Li, Zaijun et al. [1] developed a novel and selective method for the determination of Ce (III) in biological samples. The chromogenic reagent used is 2,6-dibromo-4-chlorocarboxy arsenazo. They reported that Ce (III) reacts with this reagent to form a 1:2 blue complex which absorbs at 630 M, in 1.0 M H_3PO_4 medium. It is mentioned that Beer's law is valid in the range of 0–0.8 μg/mL of Ce (III). The value of molar absorptivity is 1.6×10^5/m/cm and Sandell's sensitivity is 0.876 ng/cm^2 with a detection limit of 2.61 ng/mL as mentioned. The interference by Zn (II) up to 40 mg, Mn (III) up to 50 mg, 30 mg of V (V) and 100 mg Al (III), 10 mg of Cu (II) 10 mg of Ni (II), 10 mg of Mg (II), 10 mg of Fe (III) as well as 12 mg of Mo (VI) are tolerated. Even interference by other rare earths can be eliminated by extracting with tri-n-butylphosphate prior to determination. The method has been claimed to be selective and sensitive as well as simple. The method has been successfully applied to Ce (III) determination in biological samples.

5.1.2 Hexaacetato calix (6) arene

Andrew Swamidoss et al. [2] proposed a spectrophotometric method for determination of Ce (III) using hexaacetato calix (6) arene as a reagent and extractant. They explain that hexaacetato calix arene in xylene was used as ligand for liquid–liquid extraction of Ce (III); quantitatively, a 1×10^{-4} M of ligand in xylene at pH 7 was used and mechanical shaking was done for 7 minutes. Further, the organic phase was stripped with 1N nitric acid, and this solution was used to determine Ce (III) at 650 nm. Tolerance studies were performed for several diverse ions, and the method was successfully applied to Ce (III) determination in electronic waste water.

5.1.3 Azo dye [N-(4-(2,3-dihydroxy-5-(3,5,7-trihydroxy-4-oxo-4H-Chromen-2-yl) diazenyl) phenylsulfonyl) acetamide] [DTPPA]

Elgendy, Khaled et al. [3] synthesized, characterized and applied a new chromogenic reagent DTPPA for the spectrophotometric determination of Ce (III) in some industrial and plant samples. The method is based on the reaction of Ce (III) with DTPPA to give a brown-colored complex in an alkaline medium using borate buffer of pH 9. The λ_{max}

https://doi.org/10.1515/9783111615004-005

reported for this complex is 560 nm. A 1:2 (M: L) complex is formed. The method is enhanced if cetylpyndinium bromide is used as a surfactant. In the Beer's law linear range of 5.0–100 µg/mL of Ce (III), the values of molar absorptivity and Sandell's sensitivity are 1.020×10^4 L/mol/cm and 0.01373 µg/cm^2. The method was used to determine Ce (III) in industrial and plant samples as well.

5.1.4 2,4-Dimethylpentan-3-one (2,4-DMP)

Jawad S. K. et al. [4] proposed a spectrophotometric determination method for Ce (III) using 2,4-DMP as a reagent. They described two processes, namely, solvation and cloud-point extraction in the presence of 1 % Triton X-100 using 2,4-DMP results in a cloud-point layer of Ce^{3+}. The complex absorbs at 295 nm. The details as described are salting out with KNO_3 at 0.6 M in the presence of 100 µg Ce^{3+}/10 mL and 1×10^{-4} M (2,4-DMP) including 0.5 mL at % Triton X-100. The solution is heated at 90 °C for 15 minutes. A 1:1, (L:M) complex is formed, and the method is successful in determining Ce (III).

5.1.5 o-Methoxy phenylthiourea

Kuchekar et al. [5] developed a solvent extraction and spectrophotometry-based method for Ce (IV) determination. o-methoxyphenylthiourea was used as a reagent. It is mentioned that a ternary complex is formed after liquid–liquid extraction from 0.05 mol/L KI aqueous media using 2.0×10^{-4} mol/L of O MePT in chloroform. The measurements were made at 510 nm. Beer's law range is up to 22.5 µg mL, with molar absorptivity as 3.38×10^3 L/mol/cm and Sandell's sensitivity as 0.041 µg/cm^2. Further, the developed method has been successfully applied to the determination of Ce (IV) for binary, ternary, soil, tap water, sea water and synthetic samples.

5.1.6 Tribromoarsenazo (TB-ASA)

Jingmei et al. [6] proposed a catalytic kinetic spectrophotometric method for the determination of cerium (IV). The method involves catalytic effect of Ce (IV) on the oxidation of a tribromoarsenazo (TB-ASA) system by potassium bromate. They been described that in 0.080 mol/L H_2SO_4 medium, the λ_{max} of Ce (IV) – (TB-ASA)–$KBrO_3$ is 510 nm. There is a good linear relationship of the amount of Ce (IV) and absorbance. The Beer's law range is 5.7×0^{-8}–5.1×10^{-7} mol/L. ng. The method has been successfully applied to determine cerium in human blood and chicken liver samples, as mentioned by the authors.

5.1.7 Malachite green

Janghel et al. [7] proposed a novel highly sensitive method for the determination of cerium using a malachite green–iodide system. The method involves the bleaching of greenish-colored malachite green dye by the action of iodine, which is released by the reaction between Ce (IV) and KI in acidic medium. The linear concentration range is 0.6–4.6 μg of Ce (IV) in volume of 25 (0.024–0.184 ppm). The reported molar absorptivity and Sandell's sensitivity values are 1.36×10^3 l/mol/cm and 0.002 μg/cm^2, respectively. The method has been successfully applied to cerium determination in silicate rocks, high purely rare earth oxides, plants, soil and material water samples.

5.1.8 Potassium ferrocyanide

Garg Anita et al. [8] reported a spectrophotometric-determination method for Ce (IV) using potassium ferrocyanide as a reagent. They reported that a greenish-yellow complex is formed between Ce (IV) and potassium ferrocyanide in dilute acetic acid. The complex is measured at 425 nm even in the presence of several other ions. Ce (IV) complex has molar absorptivity as 6.0×10^3 L/mol/cm, and Sandell's sensitivity is 0.023 μg/cm^2.

5.1.9 DBC-chlorophosphanazo

Zhai, Qumgzhou et al. [9] proposed a spectrophotometric method using a color-fading reaction of DBC–Chlorophosphonazo as a reagent. They mentioned that in a medium of 1.0 mol/L ~ 2.0 mol/L sulfuric acid, the reaction of Ce (IV) with DBC–Chlorophosphonazo results into color fading. This is used to determine Ce (IV) at 530 nm. The molar absorptivity of 4.89×10^4 L/mol/cm, and the linear concentration range is 0.020 μg/mL ~0.88 μg/mL of Ce (IV). The method has been successfully applied for cerium determination in molecular sieves.

5.1.10 DBC-arsenazo

Zhai et al. [10] have proposed a spectrophotometric-determination method for cerium using fading reaction with DBC–arsenazo. They reported that in the medium of 0.18–1.08 mol/L of sulfuric acid, Ce (IV) experiences a fading reaction with DBC–arsenazo which is used for the determination of cerium (IV) using spectrophotometry. The λ_{max} is 530 nm, and the molar absorptivity of the complex is 1.03×10^4 L/mol/cm in the Beer's law range of 1.20–12.0 μg/ml of Ce (IV). Interference of about 36 coexisting ions was also studied. The method can be used to accurately determine cerium in water samples with selectivity.

5.1.11 2,6-Dibromo-4-chlorocarboxy arsenazo

Li, Zaijun et al. [11] developed a method for cerium (III) determination using 2,6-dibromo-4-chlorocarboxyarsenazo as an analytical reagent. The method involves the reaction of Ce (III) and the reagent in 1.0 M H_3PO_4 to form a 1:2 blue complex which absorbs at 630 nm. In the Beer's law range of 0–0.8 µg/mL of Ce (III), the complex is stable for 24 hours at room temperature. The reported values of molar absorptivity, Sandell's sensitivity and detection limits are 1.6×10^5 m/cm, 0.876 ng/cm^2 and 2.61 ng/mL, respectively. By prior extraction using tri-n-butyl phosphate effect of coexisting rare earths can be eliminated, and the method reportedly tolerates several ions such as 40 mg of Zn (I), 50 mg of Mn (II), 30 mg of V (II), 100 mg of Al (III), 10 mg of Cu (II), 10 mg of Ni (II), 10 mg of Mg (II) and 10 mg of Fe (III) 12 mg of Mo (VI), to mention some. The method is simple, selective and useful for the determination of Ce (III) in biological samples.

5.1.12 Leuco xylene cyanol FF

Revanasiddappa Hosakere et al. [12] proposed a method for the spectrophotometric determination of cerium with Leuco xylene cyanol FF as a reagent.

5.1.13 Phenothiazine derivatives

Masti, Saraswati P. et al. [13] proposed a spectrophotometric method for the determination of cerium (IV) using the phenothiazine derivative propionyl promazene phosphate (PPP) as a reagent. A red-colored radical cation is produced with Ce (IV) in phosphoric acid medium with λ_{max} at 513 nm. In the concentration range of 1–11 µg/mL of cerium, the method gives a Sandell's sensitivity of 16.14 ng/cm^2. Furthermore, two more phenothiazines, namely, butaperazine dimaleate and propericiazine, were also used. The method has been successfully applied to the determination of cerium in analysis of magnesium-based cerium alloys and synthetic cerium mixtures.

5.1.14 Rhodamine 6G

Vijayalekshmy et al. [14] proposed a spectrophotometric determination method for cerium (IV) using rhodamine 6 G as a reagent. The complex between the two is formed in a sulfuric acid medium that absorbs at 525 nm. The addition of Ce (IV) instantly decreases at 525, and the decrease is used to measure Ce (IV). The method is applicable to cerium determination in seawater.

5.1.15 Gallocyanin

Ensafi et al. [15] developed a spectrophotometric determination method for cerium based on ion-exchange separation coupled with a flow injection system. The discoloration of gallocyanin due to its oxidation was monitored spectrophotometrically at 524 nm. The cerium catalyzes the oxidation, and this method involves an ion-exchange separation system in the carrier stream. The detection limit is 0.25 µg and 0.30–10.0 µg is the linear range. Cerium in real samples has been determined by this method.

5.1.16 Tribromoarsenazo

Wu et al. [16] proposed a direct spectrophotometric method for the determination of cerium sub group rare earths using tribromoarsenazo. Cerium reacts with the reagent in a strong acidic medium of HCl to give a violet complex. The complex absorbs at 635 nm P, and a 1:3 complex is formed. The reported molar absorptivity is $1.15–1.25\times10^5$ L/mol/cm for the sub group (cerium) members, in the Beer's law range 0–25 µg/50 mL.

5.1.17 Butyl rhodamine B

Wang Jialin et al. [17] proposed a method of cerium and scandium determination using butyl rhodamine B. The method is based on the formation of ion association complexes of heteropoly molybdoceric acid or heteropoly molybdoceric scandic acid with butyl rhodamine B. The complex absorbs at 570 nm. The reported molar absorptivity is 3.96×10^6 L/mol/cm. Beer's law is obeyed in the range 0–04 µg/L of cerium. The detection limit is 1.0 µg/L for a 1:6:3 [Ce: Mo: BRB] complex. The method has been applied to geological samples with satisfactory results.

5.1.18 o-Tolidine

Chao [18] developed a spectrophotometric method for the determination of cerium (IV) using o-tolidine as a reagent. The method uses Ce (IV)-o-tolidine system for the determination of Ce (IV). A yellow complex is formed which absorbs at 440 nm and has a molar absorptivity of 2.5×10^4 L/mol/cm in the Beer's law range of 0–6 µg/mL. The method has been applied to trace determination of cerium in sulphuric acid with satisfactory results.

5.1.19 Chlorophosphanazo-p-benzoyl-glycine (CPA-P-BG-CPB)

Wang et al. [19] developed a spectrophotometric determination method for cerium using the reagent in the presence of alcohol. A blue complex is formed in $0.16–0.32\times10^{-3}$ mol/L of HCl medium. The reported molar absorptivity coefficient is 1.65×10^5 L/mol/cm.

5.1.20 4-Sulpho-2-amino benzenethiol

Chakrabarti [20] developed a spectrophotometric determination method for Ce (IV) in aqueous medium using 4-sulpho-2-aminobenzenethiol as a reagent. In 0.4–2.88 M HCl, a blue–violet complex is formed that absorbs at 560 nm. The Beer's law range is 0.4–18 ppm of Ce (IV). The reported values of molar absorptivity and Sandell's sensitivity are 8.06×10^3 L/mol/cm and 0.0172 μg/cm^2, respectively.

5.1.21 1-(2-pyridylazo)-2-naphthol

Abu-Zuhri [21] developed a method for cerium (III) using PAN as a reagent. They described that cerium (III) reacts with PAN at pH 10.2 in 40 % ethanol to give a red complex having a λ_{max} at 545 nm. The reported molar absorptivity is 3.95×10^3 mol/L/cm. The method has been claimed to be sensitive.

5.1.22 4-Benzoyl-2,4-dihydro-5-methyl-2-phenyl-3H-pyrazol-3-one

Arora et al. [22] proposed a spectrophotometric method for the determination of Ce (IV) using 4-benzoyl-2,4-dihydro-5-methyl-2-phenyl-3H-pyrazol-3-one as a reagent. The Ce (IV) complex with the said reagent is quantitatively extracted using 4 moles of ligand per mole of metal at pH 4.8. The method has a Sandell's sensitivity value of 0.047 μg/cm^2.

5.1.23 5-Sulfo-4-methyl-salicylic acid

Abbasi et al. [23] proposed a highly sensitive method for the determination of cerium in contaminated waters. A 0.025 M of solution is needed for full development of color. The Beer's law range is 1–15 ppm. The authors claims that the method is successful for cerium determination even in the presence of several interfering ions.

5.1.24 Promazine hydrochloride

Gowda, H. Sauke et al. [24] used promazine hydrochloride as a reagent for the spectrophotometric determination of Ce (IV). The reagent instantly forms a red-colored radical with Ce (IV) in a 0.5–2M H_2SO_4 or 0.5–2.5 M H_3PO_4 solution. This radical absorbs at 505 nm in the Beer's law range 0.5–15 ppm in H_2SO_4 or 0.5–2.1 ppm H_3PO_4. Sandell's sensitivity for both media are 0.022 and 0.019 μg/cm^2, respectively.

5.2 Spectrophotometric methods for determination of miscellaneous rare earths

5.2.1 Erbium

5.2.1.1 Kojic acid dye

Abdou et al. [25] developed a spectrophotometric method for the determination or erbium using Kojic acid dye (Koj) and cetylpyridinium chloride (CPC) as a cationic surfactant from Egyptian monazite and xenotime concentrates using third- derivative spectrophotometry.

5.2.1.2 Murexide

Turkoglu Orhan et al. [26] developed a sensitive, facile and simple method for the determination of trace amounts of erbium based on erbium–murexide complex formation. In the Beer's law linear range of 0.1–5.0 mg/L of erbium, the complex absorbs at 463 nm. The authors claimed that the method can successfully determine erbium in synthetic samples.

5.2.1.3 2-(5-Bromo-2-pyridylazo)-5-diethylaminophenol

D'Angelo et al. [27] proposed a sensitive spectrophotometric method for the determination of erbium (III) in geological materials and glass samples using (5-Br-PADAP) as a reagent. The reagent has been used in Beer's law range of 0.2–1.30 mg/L of erbium at 581 nm. The pH range is 9.0–10.9, and the molar absorptivity value of 1.36×10^3 in the presence of Triton X-100 has been reported. The method has been successful in the determination of erbium in geological materials and glass samples.

5.2.1.4 2 (3,5-dichloro-2-pyndylazo)-5-dimethylaminophenol

Silva, M. Ferranda et al. [28] used this reagent for the determination of erbium by cloud-point extraction and preconcentration of the complex. The complex Er (III)–2 (3,5-dichloro-2-pyridylazo)-5-dimethylaminophenol with non-ionic surfactant polyethylene glycolmono-p-nonyl-phenylether (PONPE 7.5) is extracted at cloud point prior to the spectrophotometric determination. The molar absorptivity is 1.27×10^5 L/mol/cm at 584 nm, in the linear range of 0.02–2 mg/L of Er (III). The lower detection limit is 1.48×10^{-7} mol/l. The method has been successful in the determination of Er (III) in synthetic samples, reproducing superconducting materials and permanent magnets.

5.2.1.5 3,5-di CIDMPAP and TX-100

Suva, Maria Fernanda et al. [29] proposed a micellar enhanced highly sensitive method for the determination of erbium (III) using 3,5-di CIDMPAP as a reagent in presence of TX-

100. The method has been used for determination of Er (III) superconducting materials and permanent magnets and has a molar absorptivity of 1.39×10^5 L/mol/cm.

5.2.2 Praseodymium

5.2.2.1 Ciprofloxacin

Wei et al. [30] proposed a spectrophotometric method for the determination of praseodymium with ciprofloxacin in normal and derivative modes A 1:3 complex is formed in the Beer's law range up to 35 μg/cm^3 with a detection limit of 0.17 μg/cm^3 of praseodymium. The authors claimed that the method yields good accuracy and selectivity.

5.2.2.2 Lomefloxacin

Wang, Naixing et al. [31] proposed a derivative spectrophotometric determination method for praseodymium with Lomefloxacin as a reagent. A 1:3 Pr $(LMFX)_3$ complex is formed in the pH 6.5–8.5 range; with a second derivative mode, the sensitivity becomes 7.4 times higher than in the normal mode. The calibration curve is linear in the range 3.6–6.5 μg/mL of Pr. The reported detection limit is 0.85 μg/mL.

5.2.2.3 5,7-Dichloroquinoline-8-ol and Rhodamine 6G

Bhagavathy et al. [32] developed a spectrophotometric method for the determination of praseodymium (m) using 5,7-dichloroquinoline-8-ol and rhodamine 6G in a weakly alkaline medium. The flotation of this complex is carried out in hexane followed by dissolution in acetone. This complex absorbs at 530 nm in the Beer's law range of 8.240 μg/L praseodymium. The resulting molar absorptivity of this complex is 1.83×10^5 l/mol/cm. It was also mentioned that praseodymium in aluminum metal and aluminum salts was determined in conjunction with an ion-exchange separation using Dowex–50W–x8 cation exchange resin.

5.2.2.4 1-(2-pyndylazo)-2-naphthol and mixture of benzophenone and naphthalene

Vyas et al. [33] proposed spectrophotometric determination method using this reagent. The extraction behavior of praseodymium with 1-(2-pyndylazo)-2-naphthol was investigated using solid, liquid and liquid–liquid extraction techniques by the authors. The method involves traces of praseodymium (III) extracted in the binary mixture of naphthalene and benzophenone. This complex is studied in the Beer's law range of 0.5–15 μg of Pr (III) with molar absorptivity and Sandell's sensitivity values of 5.4×10^4 l/mol/cm and 0.0026 μg/cm^2, respectively.

5.2.3 Yttrium

5.2.3.1 2-Hydroxy-1-naphtholdehyde-p-hydroxybenzoichydrazone (HNHBH)

Govinda Chaudhary [34] proposed a direct spectrophotometric determination method for yttrium using HNHBH as a reagent. The proposed method is both direct and derivative. The reagent reacts with yttrium in basic medium (pH 8.5 ammonium chloride and ammonium hydroxide buffer) to yield a yellow 1:4 (Y-HNHB) complex that absorbs at 410. An instantaneous color reaction occurs, resulting in a stable color. The reported value of molar absorptivity is 6.76×10^4 L/mol/cm, whereas Sandell's sensitivity value is 0.0013 µg/cm^2. The linear range of determination is 0.044–2.222 µg/mL of Y (III). Second and third derivative spectrophotometric methods have also been developed with greater sensitivity and selectivity. The method as reported has been applied to yttrium (III) determination in nickel-based alloy samples as well as geological samples.

5.2.3.2 5,7-diodoquinoline-8-ol and Rhodamine 6G

Vijayalekshmy et al. [35] proposed a spectrophotometric method for the determination of yttrium using 5,7-diodoquinoline-8-ol and rhodamine 6G as a reagent, in Y-Al alloys. The ternary ion association complex Y-5,7-diiodoquinoline-8-ol rhodamine 6G is extracted in toluene. The complex absorbs at 540 nm and the Beer's law range is 0–20 µg of yttrium. The molar absorptivity values 5.3×10^5 L/mol/cm at 540 nm.

5.2.3.3 Dibenzo-18-crown-6

Agrawal et al. [36] developed a solvent extraction method for the separation of yttrium using dibenzo-18, crown-6 followed by its spectrophotometric determination. The dibenzo-18-crown-6 [Y (OH), [L] complex is extracted in dichloromethane and absorbs at 280 nm. The molar absorptivity is 1.35×10^4 l/mol/cm, and the detection limit is 1 mg/mL. The color was developed with xylenol orange also having λ_{max} 555 and the molar absorptivity was 5.48×10^3 l/mol/cm. The system obeys Beer's law in 0.81–16.2 µg/mL of yttrium. The method is successful in the determination of yttrium in standard rock samples.

5.2.3.4 Molybdoyttric acid and basic dyes

Cheng et al. [37] proposed a method of spectrophotometric determination using the color reaction of molybdoytric acid and basic dyes in aqueous solution in the presence of poly (vinyl alcohol). It was based on ion association complex of molybdoyttric acid and three basic dyes rhodomine B (RB), butyl rhodamine B (BRB) as well as Nile Blue (NB) in an aqueous acidic solution containing poly (vinyl alcohol) PVA. The λ_{max} for each dye RB, BRB and NB are 590, 580 and 590 nm respectively and resulting molar absorptivity of the complexes are 3.53×10^5, 4.74×10^5 and 4.99×10^5 L/mol/cm in the Beer's law range up to 3.5, 3.0 and 3.5 µg of yttrium/25 mL. The method has been successfully applied to the trace determination of yttrium in magnesium alloys and some geological samples.

5.2.3.5 1-Amino-4-hydroxyanthraquinone (AMHA)

Saleh [38] proposed a spectrophotometric determination method for yttrium using AMHA as a reagent. A violet YLH (+) complex absorbs at 580 at 5.2 giving a molar absorptivity of 0.98×10^4 L/mol/cm.

5.2.3.6 4-(2-Thiazolylazo) resorcinol

Gaokar et al. [39] reported a determination method for yttrium using 4-(2-thiazolylazo) resorcinol as a reagent. A 1:3 red-colored complex is formed in the pH range 7.9–8.4 that absorbs at 540 nm. Beer's law is obeyed over the concentration range 0.12–1.4 µg of yttrium/mL. The molar absorptivity value was 6.04×10^4 l/mol/cm, and Sandell's sensitivity was reported as 1.5 ng/cm^2.

5.2.4 Thorium

5.2.4.1 Pyrimidine azo dyes and cetylpyndimium chloride

Amin, Alaa S. et al. [40] proposed a simultaneous spectrophotometric determination method for thorium and other rare earths using pynmidine azo dyes and cetylpyndimium chloride. It found that reagents 5-(2',4'-dimethyl phenylazo) 6-hydroxypynmidine-2,4-dione (1) and 5-(4'-nitro-2', 6'-dichlorophenylazo) 6-hydroxypynmidine-2,4-dione (II) in the absence of cetyl pyndinium chloride (CPC) react with thorium to yield a red complex. The molar absorptivity and Sandell' sensitivities were calculated in the absence of CPC. A simple method was proposed on this basis.

5.2.5 Holmium

5.2.5.1 2-(Diphenylacetyl) indan-1,3-dione and octylphenyl poly (ethylene glycol) ether

Wang, Nai-Xing et al. [41] proposed a zero-order and second-order derivative spectrophotometric determination method for holmium with this reagent. The λ_{max} for zero order and derivative spectra are 450 and 460 nm. The sensitivity for holmium is increased 5 times. The detection limit is 0.37 µg/mL of holmium. The method has been proven successful in the determination of holmium in synthetic samples and rare earths.

5.2.5.2 2-(5-Bromo-2-pyridylazo)-5-diethylaminophenol

Fernandez et al. [42] proposed a sensitive spectrophotometric method for the determination of holmium in geological samples using 2-(5-bromo-2-pyridylazo)-5-diethylaminophenol as a reagent. It reported reaction between Ho (III) and reagent complex is used for the determination of holmium in the concentration range 0.2–1.35 mg/L at 580 nm. The pH was optimal between 9.1 to 10.6, and molar absorptivity was reported as $1.29 \pm 0.04 \times 10^5$ l/mol/cm. The solubilizing agent Tx-100 was used for making the complex soluble in water. Using this method, it is related that Ho (III) in geological samples can be successfully determined.

5.2.6 Europium

5.2.6.1 Arsenazo III and xylenol orange

Pillai, Jisha S. et al. [43] developed a spectrophotometric method for the determination of Eu (III) using arsenazo III and xylenol orange as complexing reagents. As described, a simple, sensitive and fast spectrophotometric method had been developed for the determination of europium with arsenazo (III) in 0.01 M HCl, 0.01 M $HClO_4$ and 0.01 M H_2SO_4 media without a buffer. However, acetate buffer of pH 6.0 was used with xylenol orange. Europium complex Eu^{3+}–arsenazo complex is 1:1 for a complex in HNO_3 and $HClO_4$, whereas a 1:2 complex is formed in HCl medium. The method is performed using Job's method.

5.2.6.2 Arsenazo III

Uhroveik, Jozef et al. [44] explored the possibility of performing the spectrophotometric determination of europium using arsenazo III as a reagent in aqueous matrix. It was detailed that the complex absorbs at 655 nm and follows Beer's law in the concentration range 0–2 mg/L of Eu (III). The molar absorptivity reported was $5.5 \pm 0.2 \times 10^4$ L/mol/cm, with a limit of detection of 20-9 μg/L. It was stated that the spectrophotometric method has been applied to the determination of europium in water and mineral samples after suitable separation.

5.2.6.3 Chlortetracycline

Liawrungratha, Saisunee et al. [45] developed a flow-injection spectrophotometric method for the determination of europium using chlortetracycline as a reagent. The method is based on the complexation between europium (III) and chlortetracycline in a tris buffer of pH 8.0. This yields a yellow-colored complex absorbing at 400 nm when 100 μl of sample containing Eu (III) and arsenazo III are injected into the merged stream of chlortetracycline, and the tris buffer solution's linear range of concentration is 0.10 to 0.60 μg/mL of europium. The reported detection of limit is 0.01 μg/mL of europium. The authors claimed that the method has been successfully applied to the determination of Eu (III) in spiked water and stream sediments with an average recovery of 99.9 and 97.5 %.

5.2.7 Neodymium

5.2.7.1 N,N'-bis (2-aminobenzoyl) ethylenediamine

Kumar, K. Girish et al. [46] used N,N'-bis (2-aminobenzoyl) ethylenediamine as a reagent for the spectrophotometric determination of neodymium (III) in pure form, natural samples and industrial effluents. It was detailed that in the pH range of 6.0–7.5 neodymium, complexes with the reagent produce a complex that absorbs at 350 nm. The linear concentration range is 3.04–13.17 µg/L. The value of the molar absorptivity was 0.214×10^4 L/mol/cm. The authors claimed that the method could be successfully applied to Nd (III) determination in various samples including natural and industrial effluents without much interference from foreign ions.

5.2.7.2 Enoxacin–cetyltrimethyl ammonium bromide

Feng Yu et al. [47] developed a derivative spectrophotometric method for neodymium determination using a enoxacin–cetyltrimethylbromide system. It was mentioned that the spectra of Nd (III) – ENX was improved by adding CTMAB to attain a 7.21-fold higher sensitivity. In the linear range of 1.2×10^{-5}–2.7×10^{-4} mol/L, Nd (III) could be determined successfully with a detection limit of 1.04×10^{-5} mol/L. Furthermore, the method has been successfully applied to the determination of Nd in mixtures of rare earth oxides.

5.2.7.3 2-(5-bromo-2-pyridylazo)-5-diethylamino naphthol (5-Br.PADAP)

l Martinez, Gasquez et al. [48] proposed a novel method with high sensitivity using 5-Br-PADAP as chromogenic reagent. At pH 2.0 the reagent reacts with Nd (III) to give a complex that absorbs at 580 nm. The molar absorptivity at 580 is 1.64×10^5 L/mol/cm in the Beer's law range 0.044–1.100 mg/L of Nd (III). The method has been improved to remove difficulties by using ion-exchange chromatography and liquid–liquid extraction techniques. Using this method synthetic samples and rock reference materials have also been analyzed.

5.2.7.4 5,7-Dichloroxine and rhodamine 6G

Bhagvathy et al. [49] developed an extraction spectrophotometric method for neodymium determination using 5,7-dichloroxine and rhodamine 6G as reagents. The method is based on extracting a ternary ion association complex. The complex formed in the range of 1–20 µg of Nd (III) in 25 ml can be determined with $\varepsilon = 1.36 \times 105$ L/mol/cm. The method has been successfully applied to determination of neodymium in aluminum metal and salt samples.

5.2.7.5 1-(2-Pyndylazo)-2-naphthol

Hernandez, Mendez et al. [50] proposed a fourth-derivative spectrophotometric method for the determination of Nd (III) in glasses. The method involves measuring peak-to-valley signals of fourth-derivatives of the absorption spectra of a Nd-PAN-Triton X-100 system. The method has a detection limit of 11 ppb.

A summary of analytical reagents for determination of Ce (III) and other rare earths has been depicted in table 5.1.

Table 5.1: List of analytical reagents used for spectrophotometric determination of Ce (III) and other rare earths.

S. No.	Name of Reagent	λ_{max}/ pH/Solvent	Molar absorptivity L/mol/cm	Sandell's sensitivity µg/cm^2	Detection limit L/ng/mL
1.	2,6-Dibromo-4-chlorocarboxyarsenazo	630 1.0 M H_3PO_4	1.6×10^5	0.876	–
2.	Hexaacetato calix (6) arene	650 7	–	–	–
3.	[N-(4-(2,3-dihydroxy-5-(3,5,7-trihydroxy-4-oxo-4H-chromen-2-yl) diazenyl) phenylsulfonyl) acetamide] [DTPPA]	560 9	7.020×10^4	0.01373	–
4.	2,4-Dimethyl pentan-3-one (2,4-DMP)	295	–	–	–
5.	o-Methoxy Phenylthiourea	318	3.38×10^3	0.041	–
6.	Tribromoarsenazo (TB-ASA)	510	–	–	5.1×10^{-8}
7.	Malachite Green		1.36×10^3	0.002	–
8.	Potassium ferrocyanide	425	6.0×10^3	0.023	–
9.	DBC – Chlorophosphonazo	530	4.89×10^4	–	–
10.	DBC – arsenazo	530	1.03×10^4	–	–
11.	2,6-Dibromo-4-chlorocarboxyarsenazo	630	1.6×10^5	0.876	2.6
12.	Leuco xylene cyanol FF	–	–	–	–
13.	Phenothiazine derivative	513	–	16.14	–
14.	Rhodamine 6G	525	–	–	–
15.	Gallocyanine	524	–	–	0.25 µg
16.	Tribromo arsenazo	–	$1.15–1.25 \times 10^5$	–	–
17.	Butyl Rhodamine B	570	3.96×10^6	–	1.0 µg/L
18.	o-Tolidine	440	2.50×10^4	–	–
19.	Chlorophosphanazo-p-benzoyl-glycine	681	1.65×10^5	–	–
20.	4-Sulpho-2-aminobenzothiol	560	8.06×10^3	0.172	–

Table 5.1 (continued)

S. No.	Name of Reagent	λ_{max}/ pH/Solvent	Molar absorptivity L/mol/cm	Sandell's sensitivity μg/cm^2	Detection limit L/ng/mL
21.	1-(2-Pyndyl-azo)-2-naphthol	545 pH 10.2	3.95×10^3	–	–
22.	4-Benzoyl-2,4-dihydro-5-methyl-2-benyl-3H-pyrazol-3-one	pH 4.8	–	0.047	–
23.	5-Sulfo-4-methyl-salicylic acid	–	–	–	–
24.	Promazinehydrochloride	0.5–2 M H_2SO_4 0.5–2.5 M H_2PO_4	–	0.0220.019	–
	Erbium				
25.	Kojic acid dye & Cetyl Pyndinium chloride	–	0.14×10^3 0.13×10^3	–	1.0 μg/mL
26.	Murexide	463	–	–	–
27.	2-(5-Bromo-2-pyridylazo)-5-diethylamino phenol	581 9–10.9	1.36×10^3	–	–
28.	2 (3,5-dichloro-2-pyridylazo-5-dimethylaminophenol	584	1.27×10^5	–	1.48×10^{-7} mol/l
29.	2 (3,5-dichloro-2-pyridylazo)-5-dimethyl aminophenol & Tx-100	–	1.39×10^5	–	–
30.	Ciprofloxacin	–	–	–	0.17 μg/cm^3
31.	Lemofloxacin	6.5–8.5	–	–	0.85 μg/mL
32.	5,7-dichloro-quinoline-8-ol and Rhodamine 6G	–	530 Weakly alkaline media	1.85×10^5	–
33.	1-(2-pyridylazo)-2-naphthol and mixture of benzophenone and naphthalene	–	–	5.4×10^4	.0026 μg/cm^2

Table 5.1 (continued)

S. No.	Name of Reagent	λ_{max}/ pH/Solvent	Molar absorptivity L/mol/cm	Sandell's sensitivity µg/cm^2	Detection limit L/ng/mL
	Yttrium				
34.	2-Hydroxy-1-naphthaldehyde-p-hydroxybenzoic hydrazone (HNABH)	410	6.76×10^4	0.0013 µg/cm^2	–
35.	5-7-diiodoquinoline-8-ol and Rhodamine 6G	540	5.3×10^5	–	–
36.	Dibenzo-18-crown-6 xylenol orange	555	5.48×10^3	–	–
37.	Molybdoyttric acid and basic dyes (Rhodamine B) Butyl rhodomine B) and Nile blue	590 580 590	3.5×10^5 4.74×10^5 4.99×10^5	1.8 2.5 ng/mL 4.8	–
38.	1-Amino-4-hydroxy anthraquinone	580 6.2	0.98×10^4	–	–
39.	4-(2-Thiazolyazo) resorcinol	540 7.9–8.4	6.04×10^4	1.5 ng/cm^2	–
40.	Pyrimidine azo dyes and cetylpyridinium chloride	563 8.5 and 9.5 Borate buffer	–	–	–
	Holmium				
41.	2-(Diphenylacetyl) indan-1,3-dione and octylphenyl poly (ethyleneglycol) ether	450 460	–	–	0.37 µg/mL
42.	2-(5-bromo-2-pyridyl-azo)-5-diethyl amino phenol	580 9.1–10.6	1.29×10^5	–	–

Table 5.1 (continued)

S. No.	Name of Reagent	λ_{max}/ pH/Solvent	Molar absorptivity L/mol/cm	Sandell's sensitivity µg/cm^2	Detection limit L/ng/mL
	Europium				
43.	Arsenazo III & Xylenol orange	0.01 M HCl 0.01 M $HClO_4$ 0.01 M H_2SO_4 pH 6 acetate buffer			
44.	Arsenazo III	655	$5.5 \pm 0.2 \times 10^4$	–	20.9 µg/L
45.	Chlorotetracycline	400 8.0			.01 µg/mL
	Neodymium				
46.	N,N'-bis (2-aminobenzol) ethylenediamine	350 6.0–7.5	214×10^4	–	–
47.	Enoxacin-cetyltrimethylammonium bromide	–	–	–	1.04× 10 mol/L
48.	2-(5-bromo-2-pyridylazo)-5-diethylaminophenol (5-Br-PADAP)	580	1.65×10^5	–	–
49.	5,7-dichloro-oxine and rhodamine 6G	–	1.36×10^5	–	–
50.	1-(2-pyridylazo)-2-naphthol				11 ppb

Bibliography

[1] Li, Z., Yang, Y., Liu, J., Pan, J., Tang, J., Novel sensitive and selective method for spectrophotometric determination of cerium (III) in biological samples using 2,6-dibromo-4-chlorocarboxyarsenazo, Analytical Letters 35(12) (2002), 1959–1975.

[2] Swamidoss, A., Mercy, C., Ramaswamy, Krishnaveni, T., Gupta, V., Hexaacetato calix (6) arene as a novel extractant in the recovery of cerium (III) by liquid-liquid extraction, Materials Today: Proceedings 21 (2020), 954–957.

[3] Elgendy, K., El-Dldamony, A., Zaky, M., El-Wahab, B. A., Spectrophotometric determination of cerium (III) in some industrial and plant samples using new synthesized azo-dye reagent: Synthesis and Characterization, Egyptian Journal of Chemistry 63(7) (2020), 2405–2417.

[4] Jawad, S. K., Kadhim, M. O., Alwan, A. S., Incorporation salvation with cloud point extraction method for separation and spectrophotometric determination of cerium (III), Oriental Journal of Chemistry 33(4) (2017), 1985–1993.

[5] Kuchekar, S. R., Naval, R. M., Han, S.-H., Solvent extraction and spectrophotometric determination of cerium (IV) by using o-methoxyphenylthio urea as an analytical reagent, Solvent Extraction Research and Development 23(1) (2016), 19–29.

[6] Jingmei, L. I., Wentao, W. E. I., Catalytic kinetic spectrophotometric determination of cerium, Journal of Rare Earths 28(3) (2010), 387–390.

[7] Janghel, E. K., Parwez, Facile spectrophotometric determination of cerium using malachite green–iodide system in geological and environmental samples, Journal of Scientific and Industrial Research 68(11) (2009), 940–944.

[8] Garg, A., Susila, K., Spectrophotometric determination of Ce (IV) using potassium ferrocyanide as an analytical reagent, Journal of the Indian Chemical Society 83(6) (2006), 626–628.

[9] Zhai, Q., Hu, W., Xiaoxia, Z., Pang, W., Color fading spectrophotometric determination of cerium (IV) with DBC-chlorophosphonazo Xiyou, Jinshu Cailiao Yu Gorgcheng, Rare Metals Materials and Engineering 34(11) (2005), 1846–1848.

[10] Zhai, Q., Zhang, X., Color fading spectrophotometric determination of cerium with DBC-arsenazo, Journal of Rare Earths 22(3) (2004), 440442.

[11] Li, Z., Yong, Y., Liu, J., Pan, J., Jian, Novel sensitive and selective method for spectrophotometric determination of cerium (III) in biological samples using 2,6-dibromo-4-chlorocarboxy arsenazo, Analytical Letters 35(12) (2002), 1959–1975.

[12] Revanasiddappa, H. D., Kiran Kumar, T. N., Spectrophotometric determination of cerium with leuco xylene cyanol FF, Analytical Sciences 18(11) (2002), 1275–1278.

[13] Masti, S. P., Seetharamappa, J., Melwanki, M. B., Motohashi, N., Spectrophotometric determination of cerium (IV) using phenothiazine derivative, Analytical Sciences 18(2) (2002), 167–169.

[14] Vijayaleshmy, Biju, Rao, Selective spectrophotometric determination of cerium (IV), Indian Journal of Chemistry, Section A. Inorganic, Physical, Theoretical and Analytical 40(12) (2001), 1365–1366.

[15] Ensafi, A. A., Zomorrodian, Catalytic spectrophotometric determination of cerium by ion exchange separation coupled to a flow injection system, Journal of Analytical Chemistry 54(5) (1999), 442–445.

[16] Wu, Pan, Li, Direct spectrophotometric determination of cerium subgroup of rare earth elements in aluminium alloys with tribromoarsenazo, Caliao Gonjeheng/Journal of Materials Engineering 11, (1997), 31–33, 30.

[17] Wang, J., Li, Z., Xu, Q., Direct spectrophotometric determination of cerium and scandium with molybdate and butyl rhodamine B., Fenxi Huaxue/Chinese Journal of Analytical Chemistry 24(3) (1996), 346–347.

[18] Chao, Spectrophotometric determination of cerium (IV)-o-tolidine system, Hedongli Gongcheng/Nuclear Power Engineering 14(3) (1993), 550–552.

[19] Wang, O., Luo, Y., Micellar solubilizing spectrophotometric determination of cerium group of rare earth using chlorophosphonazo-p-benzoyl-glycine, Xiyou Jinshu/Chinese Journal of Rare Metals 6(1) (1987), 67–70.

[20] Chakrabarti, A. K., Spectrophotometric determination of cerium (V) in aqueous medium with 4-sulpho-2-aminobenzenethiol and its studies, Journal of the Indian Chemical Society 62(7) (1985), 541–543.
[21] Abu-Zuhri, A. Z., Spectrophotometric studies and analytical application of Ce (III) chelates with 1-(2-pyridylazo)-2-naphthol, Montasheffe fur Chemie 115 (1984), 57–64.
[22] Arora, H. C., Venkateswarlu, C., Rao, G. N., Extraction and spectrophotometric determination of cerium (IV) with 4-benzoyl-2,4-dihydro-5-methyl-2-phenyl-3H-pyrazol-3-one, Indian Journal of Chemistry, Section B: Inorganic, Physical, Theoretical and Analytical 19(5) (1980), 500–501.
[23] Abbasi, A., Micro determination of cobalt, cerium in contaminated waters, Indian Journal of Environmental Health 22(4) (1980), 296–301.
[24] Gowda, H. S., Thimmaiah, Spectrophotometric determination of cerium (IV), arsenic (III) and nitric acid with promazine hydrochloride, Microchemical Journal 23(3) (1978), 291–296.
[25] Abdou, A.-E.-E., Fattah, E. D., Spectrophotometric determination of erbium using Kojic acid dye in different rare earth concentrates, Egyptian Journal of Petroleum 27(4) (2018), 739–745.
[26] Turkoglu, O., Soylak, M., Colak, H., Spectrophotometric determination of erbium (III) with nurexide in synthetic samples, Asian Journal of Chemistry 14(1) (2002), 363–366.
[27] D'Angelo, Fernandez, Martinez, Marchevsky, Sensitive spectrophotometric determination of erbium (III) in geological materials and glasses with 2-(5-bromo-2-pyridylazo)-5-diethylaminophenol, Journal of Analytical Chemistry 54(6) (1999), 586–590.
[28] Silva, F. M., Fernandez, L., Olsina, R. A., Stacchiola, D., Cloud point extraction, preconcentration and spectrophotometric determination of erbium (III)-2(3,5-dichloro-2-pyridylazo)-5-dimethylaminophenol, Analytica Chimica Acta 342(2–3) (1997), 229–238.
[29] Suva, M. F., Fernandez, L., Olsina, R. A., Micellar enhanced highly sensitive spectrophotometric determination of erbium (III) with 3,5-di CIDMPAP and Tx-100, Anales de Quimica 92(6) (1996), 344–348.
[30] Wei, Naixing, Quanjie, Zhikun, Xiuqin, Fuxiang, Bo, Study of the absorption spectra of the 4f electron transition of the praseodymium complex with ciprofloxacin and its analytical application, Analyst 126(8) (2001), 1440–1443.
[31] Wang, N., Ren, X., Si, Z., Jiang, W., Liu, C., Liu, X., Derivative spectrophotometric determination of praseodymium in rare earth mixtures with Lomefloxacin, Talanta 51(3) (2000), 595–598.
[32] Bhagvathy, Prasada Rao, Damodaran, Flotation–spectrophotometric determination of praseodymium with 5,7-dichloroquinoline-8-ol and Rhodamine 6G, Analytica Chimica Acta 280(1) (1993), 169–172.
[33] Vyas, P. S., Mittal, V. P., Spectrophotometric determination of praseodymium (III) as its 1-(2-pyridylazo)-2-naphthol complex with a mixture of benzophenone and naphthalene, Indian Journal of Chemistry, Section A: Inorganic, Physical, Theoretical and Analytical 29(6) (1990), 610–611.
[34] Chowdary, G., Basha, S., Direct and derivative spectrophotometric determination of yttrium (III) using 2-hydroxy-1-naphthaldehyde-p-hydroxy benzoic hydrazone (HNHBH) in alloy and geological samples, De Pharmacia Lettre 6(6) (2014), 373–379.
[35] Vijayalekshmy, Biju, Rao, P., Spectrophotometric determination of yttrium in y-Al alloys with 5,7-diiodoquinoline-8-ol and Rhodamine 6G, Indian Journal of Chemical Technology 10(5) (2003), 466–469.
[36] Agrawal, Menon, Sudhakar, Solvent extraction and separation of yttrium with dibenzo-18-crown-6, Separation and Purification Technology 24(1) (2001), 197–203.
[37] Cheng, G., Li, Z., Wang, J., Xu, Q., Spectrophotometric determination of yttrium with molybdoyttric acid and basic dyes, Analytical Letters 30(10) (1997), 1939–1949.
[38] Salach, M. S., Spectrophotometric determination of micro amounts of yttrium with 1-amino-4-hydroxy anthraquinone, Montasheffe fur Cheme 126(6/7) (1995), 621–630.
[39] Gaokar, et al., Rapid spectrophotometric determination of yttrium (III) with 4-(2-thiazoylazo) resorcinol, Indian Journal of Chemistry, Section A: Physical, Theoretical and Analytical 25(8) (1986), 791–792.

[40] Amin, A. S., Mohammad, T. Y., Simultaneous spectrophotometric determination of thorium and rare earth metals with Pynamidine azo dyes and Cetylpyndinium chloride, Talanta 54(1) (2001), 611–620.

[41] Wang, N.-X., Jiang, W., Si, Z.-K., Xang, J.-H., Liang, W.-An., Derivative spectrophotometric determination of holmium in rare earth mixtures with 2-(diphenyl-acetyl) indan-1,3-dione and octyl-phenyl poly (ethyleneglycol) either, Mikrochimica Acta 126(3) (1997), 251–255.

[42] Fernandez, Martinez, Perino, Marchevsky, Sensitive spectrophotometric determination of holmium (III) in geological samples with 2-(5-bromo-2-pyndylazo)-5-diethylaminophenol, Anale des La Association Quimica Argentina 87(1–2) (1999), 15–24.

[43] Pillai, J. S., Chaudhary, S., Sengupta, A., Spectrophotometric determination of europium in aqueous samples and interference of Fe^{3+}, UO_2^{2+} and Th^{4+} and other complexing agents on determination, Journal of Radioanalytical and Nuclear Chemistry 331(5) (2022), 2125–2133.

[44] Uhroveik, J., Geyvathova, M., Lesny, J., Possibility of the spectrophotometric determination of europium by means of Arsenazo III, Nova Biotechnologica et Chimica 12(2) (2013), 93–99.

[45] Liawruangratha, S., Singto, S. K., Flow injection spectrophotometric determination of europium using chlortetracycline, Talanta 59(1) (2003), 9–18.

[46] Kumar, K.G, Muthuselvi, N,N'-bis (2-ammobenzoyl) ethylenediamine as a reagent for the spectrophotometric determination of neodymium (III), Asian Journal of Chemistry 23(8) (2011), 3620–3622.

[47] Yu, F., Jiang, W., Wang, N., Zhao, W., Derivative spectrophotometric determination of neodymium with neodymium-enoxacin – cetyltrimethylammonium bromide system, Fenxi Huaxue Chinese Journal of Analytical Chemistry 31(1) (2003), 92–94.

[48] Martinez, Gasquez, Olsina, Marchevsky, Spectrophotometric determination of neodymium in high sensitivity conditions in rocks with 2-(5-bromo-2-pyridylazo)-5-diethylaminophenol (5-Br-PADAP), Chemica Analytical CZNA 41(2) (1996), 275–282.

[49] Bhagvathy, V., Reddy, M. L. P., Rao, T. P., Damodaran, A. D., Extractive spectrophotometric determination of neodymium with 5,7-dichloroxine and rhodamine 6G, Indian Journal of Chemistry, Section A: Inorganic, Physical, Theoretical and Analytical 32(5) (1993), 463–464.

[50] Mendez, H., Cordeo, M., Pavon, P., Miraues, C., Determination of neodymium with 1-(2-pyridylazo)-1-naphthol by high order derivative spectrophotometry. Application to the determination of neodymium in glasses, Analyst 113 (1988), 429–431.

Index

https://doi.org/10.1515/9783111615004-006

www.ingramcontent.com/pod-product-compliance
Lightning Source LLC
LaVergne TN
LVHW082258150926
848178LV00005B/85